THE BEE-MASTER OF WARRILOW

THE BEE-MASTER OF WARRILOW

BY

TICKNER EDWARDES

FELLOW OF THE ENTOMOLOGICAL SOCIETY OF LONDON

AUTHOR OF "THE LORE OF THE HONEY-BEE"

WITH A FOREWORD BY

BERNARD PRICE

NEW AND ENLARGED EDITION

ASHGROVE PRESS, BATH

This edition published in Great Britain by
ASHGROVE PRESS LIMITED
26 Gay Street, Bath, Avon BA1 2PD

Enlarged edition first published by
Methuen & Co Limited, 1920

ISBN 0 906798 33 7

Printed in Great Britain by
Hillman Printers (Frome Limited)

FOREWORD

'WARRILOW' wrote Tickner Edwardes, 'is a precipitous village tucked away under the green brink of the Sussex Downs'. The real name of that village is Burpham, perched like a fortress above the tidal and fast flowing River Arun. Indeed, it was once a fortified place, and an ancient earthwork clearly proclaims it.

The Bee-Master of Warrilow was the first book that Tickner Edwardes wrote on the subject of bees; two earlier works being *Sidelights of Nature*, which appeared in 1898, and the delightful *An Idler in the Wilds* published 1906. He went on to publish eighteen books in all, including novels and the far famed *The Lore of the Honey Bee*. It was, however, *The Bee-Master of Warrilow* that first made his name. The book is a fascinating blend of the bee-lore of the past and the advent of modern apiculture. Tickner Edwardes has gleaned much from the bee-masters of the Victorian era and, what is so delightful for the reader, is the manner in which he has delved among the knowledge and memories of the old bee-masters, and also retained for us their lines of thought and the patterns of their speech.

Tickner Edwardes was born in London in 1865 and his love of the countryside developed from boyhood. He worked as a journalist until he had sufficient funds to enable him to settle at Burpham, just a few miles from Arundel, with its great castle. He and his family first lived in a small thatched cottage called The Den, since demolished, but the old flint garden wall still remains; it was here that his early books were written.

With the approach of his fiftieth birthday, Tickner joined the Royal Army Medical Corps as a private in 1915, and served in Gallipoli. By January 1919 he had attained the rank of captain and become a medical entomologist on the Staff of Colonel Sir Ronald Ross, malaria consultant to the War Office. After the war he took Holy Orders, becoming Vicar of Burpham in 1927, retiring in 1935.

Tickner Edwardes is still well remembered in the village, and further afield, today. Those who knew him recall his strong voice and gentle manner, and his shy brown eyes. Some people have described him as bearing a remarkable likeness to George Arliss, the early film star. *Tansy* his second novel was also produced as a silent film and was shown in cinemas throughout the world during the 1920s.

Yet the overwhelming interest of Tickner Edwardes centred itself around bees. He was a man who noticed little things, and he once wrote: 'The commonest things in Nature are always the most beautiful. To link rarity with loveliness is but the human infraction of the wider theme.' Most of his books contain at least some mention of bees. How many people today have ever heard of the 'ringing of bees' when the old bee men might beat upon a saucepan with a heavy spoon or a key while a swarm of bees flew wildly around them. The 'ringing' stopped only when the swarm began to settle in a tree, to be shaken later into a straw hive or 'skep' as they were called. Tickner well knew the fondness that existed for the straw hive, and the suspicion with which his own 'boxes' were frequently eyed.

It must not be thought that Tickner merely recorded old methods of bee-keeping and the transition to the new, for his scholarly research into the subject reached back to the writings of Virgil some two thousand years ago, and even to the days of the Pharaohs. Just as Virgil had retired among lemon groves and bee hives, so too did Tickner, although apples were substituted for lemons! At Burpham he contemplated the lives of bees in terms of a stern political economy, and he recognised that beekeeping must have been among the earliest occupations of man; and that a bee among Egyptian hieroglyphics symbolised royalty.

Within these pages we are also reminded how bees have long been used as physicians said to cure rheumatism by their string, and all manner of ailments with the sweetness of their honey.

I believe that all who read *The Bee-Master of Warrilow* will cause them to look at bees in a new light. Meanwhile, the descendants of the hives of the Rev. Tickner Edwardes, still gather their pollen among the flower gardens and the woods and hills of Burpham.

BERNARD PRICE

PREFACE TO NEW EDITION

THE original " BEE-MASTER OF WARRILOW "—
that queer little honey-coloured book of far-
off days—contained but eleven chapters: in its
present edition the book has grown to more than
three times its former length, and constitutes practically a new volume.

To those who knew and loved the old " BEE-
MASTER OF WARRILOW," no apology for the
additional chapters will be required, because it is
directly to the solicitation of many of them that this
larger collection of essays on English bee-garden
life owes its appearance. And equally, to those who
will make the old bee-man's acquaintance for the
first time in these present pages, little need be said.
In spite of the War, the honey-bee remains the same
mysterious, fascinating creature that she has ever
been; and the men who live by the fruit of her toil
share with her the like changeless quality. The
Master of Warrilow and his bees can very well be
left to win their own way into the hearts of new
readers as they did with the old.

T. E.

THE RED COTTAGE,
   BURPHAM, ARUNDEL,
      SUSSEX.

CONTENTS

DEDICATION

INTRODUCTION

AMONG the beautiful things of the country-side, which are slowly but surely passing away, must be reckoned the old Bee Gardens—fragrant, sunny nooks of blossom, where the bees are housed only in the ancient straw skeps, and have their own way in everything, the work of the bee-keeper being little more than a placid looking-on at events of which it would have been heresy to doubt the finite perfection.

To say, however, that modern ideas of progress in bee-farming must inevitably rob the pursuit of all its old-world poetry and picturesqueness, would be to represent the case in an unnecessarily bad light. The latter-day beehive, it is true, has little more æsthetic value than a Brighton bathing-machine; and the new class of bee-keepers, which is springing up all over the country, is composed mainly of people who have taken to the calling as they would to any other lucrative business, having, for the most part, nothing but a good-humoured contempt alike for the old-fashioned bee-keeper and the ancient traditions and superstitions of his craft.

Nor can the inveterate, old-time skeppist himself —the man who obstinately shuts his eyes to all that is good and true in modern bee-science—be counted on to help in the preservation of the beautiful old gardens, or in keeping alive customs which have been handed down from generation to generation, almost unaltered, for literally thousands of years. Here and there, in the remoter parts of the country, men can still be found who keep their bees much in the same way as bees were kept in the time of Columella or Virgil; and are content with as little profit. But these form a rapidly diminishing class. The advantages of modern methods are too over-whelmingly apparent. The old school must choose between the adoption of latter-day systems, or suffer the only alternative—that of total extinction at no very distant date.

Luckily for English bee-keeping, there is a third class upon which the hopes of all who love the ancient ways and days, and yet recognise the absorbing interest and value of modern research in apiarian science, may legitimately rely. Born and bred amongst the hives, and steeped from their earliest years in the lore of their skeppist forefathers, these interesting folk seem, nevertheless, imbued to the core with the very spirit of progress. While retaining an unlimited affection for all the quaint old methods in bee-keeping, they maintain them-selves, unostentatiously, but very thoroughly, abreast of the times. Nothing new is talked of in the world of bees that these people do not make trial of, and quietly adopt into their daily practice, if really serviceable; or as quietly discard, if the contrivance prove to have little else than novelty to recommend it.

As a rule, they are reserved, silent men, difficult of approach; and yet, when once on terms of familiarity, they make the most charming of companions. Then they are ever ready to talk about their bees, or discuss the latest improvements in apiculture; to explain the intricacies of bee-life, as revealed by the foremost modern observers, or to dilate by the hour on the astounding delusions of mediæval times. But they all seem to possess one invariable characteristic—that of whole-hearted reverence for the customs of their immediate ancestors, their own fathers and grandfathers. In a long acquaintance with bee-men of this class, I have never yet met with one who could be trapped into any decided admission of defect in the old methods, which—to say truth—were often as senseless as they were futile, even when not directly contrary to the interest of the bee-owner, or the plain, obvious dictates of humanity. In this they form a refreshing contrast to the ultra-modern, pushing young apiculturist of to-day; and it is as a type of this class that the Bee-Master of Warrilow is presented to the reader.

THE BEE-MASTER OF WARRILOW

CHAPTER I

THE BEE-MASTER OF WARRILOW

LONG, lithe, and sinewy, with three score years of sunburn on his keen, gnarled face, and the sure stride of a mountain goat, the Bee-Master of Warrilow struck you at once as a notable figure in any company.

Warrilow is a little precipitous village tucked away under the green brink of the Sussex Downs; and the bee-farm lay on the southern slope of the hill, with a sheltering barrier of pine above, in which, all day long, the winter wind kept up an impotent complaining. But below, among the hives, nothing stirred in the frosty, sun-riddled air. Now and again a solitary worker-bee darted up from a hive door, took a brisk turn or two in the dazzling light, then hurried home again to the warm cluster. But the flash and quiver of wings, and the drowsy song of summer days, were gone in the iron-bound January weather; and the bee-master was lounging idly to and fro in the great main-way of the waxen city, shot-gun under arm, and with apparently nothing more to do than to meditate over past achievements, or to plan out operations for the season to come.

As I approached, the sharp report of the gun rang out, and a little cloud of birds went chippering fearsomely away over the hedgerow. The old man watched them as they flew off dark against the snowy hillside. He threw out the cartridge-cases disgustedly.

" Blue-tits!" said he. " They are the great pest of the bee-keeper in winter time. When the snow covers the ground, and the frost has driven all insect-life deep into the crevices of the trees, all the blue-caps for miles round trek to the bee-gardens. Of course, if the bees would only keep indoors they would be safe enough. But the same cause that drives the birds in lures the bees out. The snow reflects the sunlight up through the hive-entrances, and they think the bright days of spring have come, and out they flock to their death. And winter is just the time when every single bee is valuable. In summer a few hundreds more or less make little difference, when in every hive young bees are maturing at the rate of several thousands a day to take the place of those that perish. But now every bee captured by the tits is an appreciable loss to the colony. They are all nurse-bees in the winter-hives, and on them depends the safe hatching-out of the first broods in the spring season. So the bee-keeper would do well to include a shot-gun among his paraphernalia, unless he is willing to feed all the starving tits of the countryside at the risk of his year's harvest."

" But the blue-cap," he went on, " is not always content to wait for his breakfast until the bees voluntarily bring it to him. He has a trick of enticing them out of the hive which is often successful even in the coldest weather. Come into the extracting-house yonder, and I may be able to show you what I mean."

He led the way to a row of outbuildings which flanked the northern boundary of the garden and formed additional shelter from the blustering gale. A window of the extracting-house overlooked the whole extent of hives. Opening this from within with as little noise as possible, the bee-master put a strong field-glass into my hand.

" Now that we are out of sight," he said, " the tits will soon be back again. There they come— whole families of them together! Now watch that green hive over there under the apple-tree."

Looking through the glass, I saw that about a dozen tits had settled in the tree. Their bright plumage contrasted vividly with the sober green and grey of the lichened boughs, as they swung themselves to and fro in the sunshine. But presently the boldest of them gave up this pretence of searching for food among the branches, and hopped down upon the alighting-board of the hive. At once two or three others followed him; and then began an ingenious piece of business. The little company fell to pecking at the hard wood with their bills, striking out a sharp ringing tattoo plainly audible even where we lay hidden. The old bee-man snorted contemptuously, and the cartridges slid home into the breech of his gun with a vicious snap.

" Now keep an eye on the hive-entrance," he said grimly.

The glass was a good one. Now I could plainly make out a movement in this direction. The noise and vibration made by the birds outside had roused the slumbering colony to a sense of danger. About a dozen bees ran out to see what it all meant, and were immediately pounced upon. And then the gun spoke over my head. It was a shot into the air, but it served its harmless purpose. From every bush and tree there came over to us a dull whirr of wings like far-off thunder, as the blue marauders sped away for the open country, filling the air with their frightened jingling note.

Perhaps of all cosy retreats from the winter blast it has ever been my good fortune to discover, the extracting-room on Warrilow bee-farm was the brightest and most comfortable. In summer-time the whole life of the apiary centred here; and the stress and bustle, inevitable during the season of the great honey-flow, obscured its manifold possibilities. But in winter the extracting-machines were, for the most part, silent; and the natural serenity and cosiness of the place reasserted themselves triumphantly. From the open furnace-door a ruddy warmth and glow enriched every nook and corner of the long building. The walls were lined with shelves where the polished tin vessels, in which the surplus honey was stored, gave back the fire-shine in a hundred flickering points of amber light. The work of hive-making in the neighbouring sheds was going briskly forward, but the noise of hammering, the shrill hum of sawing and planing machinery, and the intermittent cough of the oil-engine reached us only as a subdued, tranquil murmur—the very voice of rest.

The bee-master closed the window behind its thick bee-proof curtains, and, putting his gun away in a corner, drew a comfortable high-backed settle near to the cheery blaze. Then he disappeared for a moment, and returned with a dusty cobweb-shrouded bottle, which he carried in a wicker cradle as a butler would bear priceless old wine. The cork came out with a ringing jubilant report, and the pale, straw-coloured liquid foamed into the glasses like champagne. It stilled at once, leaving the whole inner surface of the glass veneered with golden bells. The old bee-man held it up critically against the light.

" The last of 19—," he said, regretfully. " The finest mead year in this part of the country for many a decade back. Most people have never tasted the old Anglo-Saxon drink that King Alfred loved, and probably Harold's men made merry with on the eve of Hastings. So they can't be expected to know that metheglin varies with each season as much as wine from the grape."

Of the goodness of the liquor there admitted no question. It had the bouquet of a ripe Ribston pippin, and the potency of East Indian sherry thrice round the Horn. But its flavour entirely eluded all attempt at comparison. There was a sugges-tive note of fine old perry about it, and a dim reminder of certain almost colourless Rhenish wines, never imported, and only to be encountered in moments of rare and happy chance. Yet neither of these parallels came within a sunbeam's length of the truth about this immaculate honey-vintage of Warrilow. Pondering over the liquor thus, the thought came to me that nothing less than a supreme occasion could have warranted its production to-day. And this conjecture was immediately verified. The bee-master raised his glass above his head.

" To the Bees of Warrilow!" he said, lapsing into the broad Sussex dialect, as he always did when much moved by his theme. " Forty-one years ago to-day the first stock I ever owned was fixed up out there under the old codlin-tree; and now there are two hundred and twenty of them. 'Twas before you were born, likely as not; and bee science has seen many changes since then. In those days there were nothing but the old straw skeps, and most bee-keepers knew as little about the inner life of their bees as we do of the bottom of the South Pacific. Now things are very different; but the improvement is mostly in the bee-keepers themselves. The bees are exactly as they always have been, and work on the same principles as they did in the time of Solomon. They go their appointed way inexorably, and all the bee-master can do is to run on ahead and smooth the path a little for them. Indeed, after forty odd years of bee-keeping, I doubt if the bees even realise that they are ' kept ' at all. The bee-master's work has little more to do with their progress than the organ-blower's with the tune."

" Can you," I asked him, as we parted, " after all these years of experience, lay down for beginners in beemanship one royal maxim of success above any other? "

He thought it over a little, the gun on his shoulder again.

"Well, they might take warning from this same King Solomon," he said, "and beware the foreign feminine element. Let British bee-keepers cease to import queen bees from Italy and elsewhere, and stick to the good old English Black. All my bees are of this strain, and mostly from one pure original Sussex stock. The English black bee is a more generous honey-maker in indifferent seasons; she does not swarm so determinedly, under proper treatment, as the Ligurians or Carniolans; and, above all, though she is not so handsome as some of her Continental rivals, she comes of a hardy northern race, and stands the ups and downs of the British winter better than any of the fantastic yellow-girdled crew from overseas."

CHAPTER II

FEBRUARY AMONGST THE HIVES

THE midday sun shone warm from a cloudless
sky. Up in the highest elm-tops the south-
west wind kept the chattering starlings gently
swinging, but below in the bee-garden scarce a
breath moved under the rich soft light.

As I lifted the latch of the garden-gate, the sharp
click brought a stooping figure erect in the midst of
the hives; and the bee-master came down the red-
tiled winding path to meet me. He carried a box
full of some yellowish powdery substance in one
hand, and a big pitcher of water in the other; and
as usual, his shirt-sleeves were tucked up to
the shoulder, baring his weather-browned arms to
the morning sun.

" When do we begin the year's bee-work? " he
said, repeating my question amusedly. " Why, we
began on New Year's morning. And last year's
work was finished on Old Year's night. If you go
with the times, every day in the year has its work
on a modern bee-farm, either indoors or out."

" But it is on these first warm days of spring,"
he continued, as I followed him into the thick of
the hives, " that outdoor work for the bee-man
starts in earnest. The bees began long ago.

24

January was not out before the first few eggs were laid right in the centre of the brood-combs. And from now on, if only we manage properly, each bee-colony will go on increasing until, in the height of the season, every queen will be laying from two thousand to three thousand eggs a day."

He stopped and set down his box and his pitcher. " If we manage properly. But there's the rub. Success in bee-keeping is all a question of numbers. The more worker-bees there are when the honey-flow begins, the greater will be the honey-harvest. The whole art of the bee-keeper consists in maintaining a steady increase in population from the first moment the queens begin to lay in January, until the end of May brings on the rush of the white clover, and every bee goes mad with work from morning to night. Of course, in countries where the climate is reasonable, and the year may be counted on to warm up steadily month by month, all this is fairly easy; but with topsy-turvy weather, such as we get in England, it is a vastly different matter. Just listen to the bees now! And this is only February!"

A deep vibrating murmur was upon the air. It came from all sides of us; it rose from under foot, where the crocuses were blooming; it seemed to fill the blue sky above with an ocean of sweet sound. The sunlight was alive with scintillating points of light, like cast handfuls of diamonds, as the bees darted hither and thither, or hovered in little joyous companies round every hive. They swept to and fro between us; gambolled about our heads; came with a sudden shrill menacing note and scrutinised our mouths, our ears, our eyes, or settled on our hands and faces, comfortably, and with no apparent haste to be gone. The bee-master noted my growing uneasiness, not to say trepidation.

" Don't be afraid," he said. " It is only their companionableness. They won't sting—at least, not if you give them their way. But now come and see what we are doing to help on the queens in their work."

At different stations in the garden I had noticed some shallow wooden trays standing among the hives. The old bee-man led the way to one of these. Here the humming was louder and busier than ever. The tray was full of fine wood-shavings, dusted over with the yellow powder from the bee-master's box; and scores of bees were at work in it, smothering themselves from head to foot, and flying off like golden millers to the hives.

" This is pea-flour," explained the master, " and it takes the place of pollen as food for the young bees, until the spring flowers open and the natural supply is available. This forms the first step in the bee-keeper's work of patching up the defective English climate. From the beginning our policy is to deceive the queens into the belief that all is prosperity and progress outside. We keep all the hives well covered up, and contract the entrances, so that a high temperature is maintained within, and the queens imagine summer is already advancing. Then they see the pea-flour coming in plentifully, and conclude that the fields and hillsides are covered with flowers; for they never come out of the hives except at swarming-time, and must judge of the year by what they see around them. Then in a week or two we shall put the spring-feeders on, and give each hive as much syrup as the bees can take down; and this, again, leads the queens into the belief that the year's food-supply has begun in earnest. The result is that the winter lethargy in the hive is soon completely overthrown, the queens begin to lay unrestrictedly, and the whole colony is forging on towards summer strength long before there is any natural reason for it."

We were stooping down, watching the bees at the nearest hive. A little cloud of them was hovering in the sunshine, heads towards the entrance, keeping up a shrill jovial contented note as they flew. Others were roving round with a vagrant, workless air, singing a low desultory song as they trifled about among the crocuses, passing from gleaming white to rich purple, then to gold, and back again to white, just as the mood took them. In the hive itself there was evidently a kind of spring-cleaning well in progress. Hundreds of the bees were bringing out minute sand-coloured particles, which accumulated on the alighting-board visibly as we watched. Now and again a worker came backing out, dragging a dead bee laboriously after her. Instantly two or three others rushed to help in the task, and between them they tumbled the carcass over the edge of the foot-board down among the grass below. Sometimes the burden was of a pure white colour, like the ghost of a bee, perfect in shape, with beady black eyes, and its colourless wings folded round it like a cere-cloth. Then it seemed to be less weighty, and its carrier usually shouldered the gruesome thing, and flew away with it high up into the sunshine, and swiftly out of view.

"Those are the undertakers," said the bee-master, ruminatively filling a pipe. "Their work is to carry the dead out of the hive. That last was one of the New Year's brood, and they often die in the cell like that, especially at the beginning of the season. All that fine drift is the cell-cappings thrown down during the winter from time to time as the stores were broached, and every warm day sees them cleaning up the hive in this way. And now watch these others—these that are coming and going straight in and out of the hive."

I followed the pointing pipe-stem. The alighting-stage was covered with a throng of bees, each busily intent on some particular task. But every now and then a bee emerged from the hive with a rush, elbowed her way excitedly through the crowd, and darted straight off into the sunshine without an instant's pause. In the same way others were returning, and as swiftly disappearing into the hive.

"Those are the water-carriers," explained the master. "Water is a constant need in bee-life almost the whole year round. It is used to soften the mixture of honey and pollen with which the young grubs and newly-hatched bees are fed; and the old bees require a lot of it to dilute their winter stores. The river is the traditional watering-place for my bees here, and in the summer it serves very well; but in the winter hundreds are lost either through cold or drowning. And so at this time we give them a water-supply close at home."

He took up his pitcher, and led the way to the other end of the garden. Here, on a bench, he showed me a long row of glass jars full of water, standing mouth downward, each on its separate plate of blue china. The water was oozing out round the edges of the jars, and scores of the bees were drinking at it side by side, like cattle at a trough.

"We give it them lukewarm," said the old bee-man, "and always mix salt with it. If we had sea-water here, nothing would be better; seaside bees often go down to the shore to drink, as you may prove for yourself on any fine day in summer. Why are all the plates blue? Bees are as fanciful in their ways as our own women-folk, and in nothing more than on the question of colour. Just this particular shade of light blue seems to attract them more than any other. Next to that, pure white is a favourite with them; but they have a pronounced dislike to anything brilliantly red, as all the old writers about bees noticed hundreds of years ago. If I were to put some of the drinking-jars on bright red saucers now, you would not see half as many bees on them as on the pale blue."

We moved on to the extracting-house, whence the master now fetched his smoker, and a curious knife, with a broad and very keen-looking blade. He packed the tin nozzle of the smoker with rolled brown paper, lighted it, and, by means of the little bellows underneath, soon blew it up into full strength. Then he went to one of the quietest hives, where only a few bees were wandering aimlessly about, and sent a dense stream of smoke into the entrance. A moment later he had taken the roof and coverings off, and was lifting out the central comb-frames one by one, with the bees clinging in thousands all about them.

"Now," he said, "we have come to what is

really the most important operation of all in the bee-keeper's work of stimulating his stocks for the coming season. Here in the centre of each comb you see the young brood; but all the cells above and around it are full of honey, still sealed over and untouched by the bees. The stock is behind time. The queen must be roused at once to her responsibilities, and here is one very simple and effective way of doing it."

He took the knife, deftly shaved off the cappings from the honey-cells of each comb, and as quickly returned the frames, dripping with honey, to the brood-nest. In a few seconds the hive was comfortably packed down again, and he was looking round for the next languid stock.

" All these slow, backward colonies," said the bee-master, as he puffed away with his smoker, " will have to be treated after the same fashion. The work must be smartly done, or you will chill the brood; but, in uncapping the stores like this, right in the centre of the brood-nest, the effect on the stock is magical. The whole hive reeks with the smell of honey, and such evidence of prosperity is irresistible. To-morrow, if you come this way, you will see all these timorous bee-folk as busy as any in the garden."

CHAPTER III

A TWENTIETH CENTURY BEE-FARMER

IT was sunny spring in the bee-garden. The thick elder-hedge to the north was full of young green leaf; everywhere the trim footways between the hives were marked by yellow bands of crocus-bloom, and daffodils just showing a golden promise of what they would be in a few warm days to come. From a distance I had caught the fresh spring song of the hives, and had seen the bee-master and his men at work in different quarters of the mimic city. But now, drawing nearer, I observed they were intent on what seemed to me a perfectly astounding enterprise. Each man held a spoon in one hand and a bowl of what I now knew to be pea-flour in the other, and I saw that they were busily engaged in filling the crocus-blossoms up to the brim with this inestimable condiment. My friend the bee-master looked up on my approach, and, as was his wont, forestalled the inevitable questioning.

"This is another way of giving it," he explained, "and the best of all in the earliest part of the season. Instinct leads the bees to the flowers for pollen-food when they will not look for it elsewhere; and as the natural supply is very meagre, we just help them in this way."

As he spoke I became rather unpleasantly aware of a change of manners on the part of his winged people. First one and then another came harping round, and, settling comfortably on my face, showed no inclination to move again. In my ignorance I was for brushing them off, but the bee-master came hurriedly to my rescue. He dislodged them with a few gentle puffs from his tobacco-pipe.

"That is always their way in the spring-time," he explained. "The warmth of the skin attracts them, and the best thing to do is to take no notice. If you had knocked them off you would probably have been stung."

"Is it true that a bee can only sting once?" I asked him, as he bent again over the crocus beds.

He laughed.

"What would be the good of a sword to a soldier," he said, "if only one blow could be struck with it? It is certainly true that the bee does not usually sting a second time, but that is only because you are too hasty with her. You brush her off before she has had time to complete her business, and the barbed sting, holding in the wound, is torn away, and the bee dies. But now watch how the thing works naturally."

A bee had settled on his hand as he was speaking. He closed his fingers gently over it, and forced it to sting.

"Now," he continued, quite unconcernedly, "look what really happens. The bee makes two or three lunges before she gets the sting fairly home. Then the poison is injected. Now watch what she does afterwards. See! she has finished her work, and is turning round and round! The barbs are arranged spirally on the sting, and she is twisting it out corkscrew-fashion. Now she is free again! there she goes, you see, weapon and all; and ready to sting again if necessary."

The crocus-filling operation was over now, and the bee-master took up his barrow and led the way to a row of hives in the sunniest part of the garden. He pulled up before the first of the hives, and lighted his smoking apparatus.

" These," he said, as he fell to work, " have not been opened since October, and it is high time we saw how things are going with them."

He drove a few strong puffs of smoke into the entrance of the hive and removed the lid. Three or four thicknesses of warm woollen quilting lay beneath. Under these a square of linen covered the tops of the frames, to which it had been firmly propolised by the bees. My friend began to peel this carefully off, beginning at one corner and using the smoker freely as the linen ripped away.

" This was a full-weight hive in the autumn," he said, " so there was no need for candy-feeding. But they must be pretty near the end of their stores now. You see how they are all together on the three or four frames in the centre of the hive? The other combs are quite empty and deserted. And look how near they are clustering to the top of the bars! Bees always feed upwards, and that means we must begin spring-feeding right away."

He turned to the barrow, on which was a large

box, lined with warm material, and containing bar frames full of sealed honeycomb.

"These are extra combs from last summer. I keep them in a warm cupboard over the stove at about the same temperature as the hive we are going to put them into. But first they must be uncapped. Have you ever seen the Bingham used?"

From the inexhaustible barrow he produced the long knife with the broad, flat blade; and, poising the frame of honeycomb vertically on his knee, he removed the sheet of cell-caps with one dexterous cut, laying the honey bare from end to end. This frame was then lowered into the hive with the uncapped side close against the clustering bees. Another comb, similarly treated, was placed on the opposite flank of the cluster. Outside each of these a second full comb was as swiftly brought into position. Then the sliding inner walls of the brood-nest were pushed up close to the frame, and the quilts and roof restored. The whole seemed the work of a few moments at the outside.

"All this early spring work," said the bee-master, as we moved to the next hive, "is based upon the recognition of one thing. In the south here the real great honey-flow comes all at once: very often the main honey-harvest for the year has to be won or lost during three short weeks of summer. The bees know this, and from the first days of spring they have only the one idea—to create an immense population, so that when the honey-flow begins there may be no lack of harvesters. But against this main idea there is another one—their ingrained and invincible caution. Not an egg will

be laid nor a grub hatched unless there is reasonable chance of subsistence for it. The populace
of the hive must be increased only in proportion
to the amount of stores coming in. With a good
spring, and the early honey plentiful, the queen will
increase her production of eggs with every day, and
the population of the hive will advance accordingly.
But if, on the very brink of the great honey-flow,
there comes, as is so often the case, a spell of cold
windy weather, laying is stopped at once; and, if
the cold continues, all hatching grubs are destroyed
and the garrison put on half-rations. And so the
work of months is undone."

He stooped to bring his friendly pipe to my
succour again, for a bee was trying to get down
my collar in the most unnerving way, and another
had apparently mistaken my mouth for the front-
door of his hive. The intruders happily driven off,
the master went back to his work and his talk
together.

"But it is just here that the art of the bee-
keeper comes in. He must prevent this interruption
to progress by maintaining the confidence of the
bees in the season. He must create an artificial
plenty until the real prosperity begins. Yet, after
all, he must never lose sight of the main principle,
of carrying out the ideas of the bees, not his own.
In good beemanship there is only one road to
success: you must study to find out what the bees
intend to do, and then help them to do it. They
call us bee-masters, but bee-servants would be much
the better name. The bees have their definite plan
of life, perfected through countless ages, and
nothing you can do will ever turn them from it.

You can delay their work, or you can even thwart it altogether, but no one has ever succeeded in changing a single principle in bee-life. And so the best bee-master is always the one who most exactly obeys the orders from the hive."

CHAPTER IV

CHLOE AMONG THE BEES

THE bee-mistress looked at my card, then put its owner under a like careful scrutiny. In the shady garden where we stood, the sunlight fell in quivering golden splashes round our feet. High overhead, in the purple elm-blossom, the bees and the glad March wind made rival music. Higher still a ripple of lark-song hung in the blue, and a score of rooks were sailing by, filling the morning with their rich, deep clamour of unrest.

The bee-mistress drew off her sting-proof gloves in thoughtful deliberation.

"If I show you the bee-farm," said she, eyeing me somewhat doubtfully, "and let you see what women have done and are doing in an ideal feminine industry, will you promise to write of us with seriousness? I mean, will you undertake to deal with the matter for what it is—a plain, business enterprise by business people—and not treat it flippantly, just because no masculine creature has had a hand in it?"

"This is an attempt," she went on—the needful assurances having been given—"an attempt, and, we believe, a real solution to a very real difficulty. There are thousands of educated women in the towns who have to earn their own bread; and they do it usually by trying to compete with men in walks of life for which they are wholly unsuited. Now, why do they not come out into the pure air and quiet of the countryside, and take up any one of several pursuits open there to a refined, well-bred woman? Everywhere the labourers are forsaking the land and crowding into the cities. That is a farmers' problem, with which, of course, women have nothing to do. The rough, heavy work in the cornfields must always be done either by men or machinery. But there are certain employments, even in the country, that women can invariably undertake better than men, and bee-keeping is one of them. The work is light. It needs just that delicacy and deftness of touch that only a woman can bring to it. It is profitable. Above all, there is nothing about it, from first to last, of an objectionable character, demanding masculine interference. In poultry-farming, good as it is for women, there must always be a stony-hearted man about the place to do unnameable necessary things in a fluffy back-shed. But bee-keeping is clean, clever, humanising, open-air work—essentially women's work all through."

She had led the way through the scented old-fashioned garden, towards a gate in the farther wall, talking as she went. Now she paused, with her hand on the latch.

" This," she said, " we call the Transition Gate. It divides our work from our play. On this side of it we have the tennis-court and the croquet, and other games that women love, young or old. But it is all serious business on the other side. And now you shall see our latter-day Eden, with its one unimportant omission."

As the door swung back to her touch, the murmur that was upon the air grew suddenly in force and volume. Looking through, I saw an old orchard, spacious, sun-riddled, carpeted with green; and, stretching away under the ancient apple-boughs, long, neat rows of hives, a hundred or more, all alive with bees, winnowing the March sunshine with their myriad wings.

Here and there in the shade-dappled pleasance figures were moving about, busily at work among the hives, figures of women clad in trim holland blouses, and wearing bee-veils, through which only a dim guess at the face beneath could be hazarded. Laughter and talk went to and fro in the sun-steeped quiet of the place; and one of the fair bee-gardeners near at hand—young and pretty, I could have sworn, although her blue gauze veil disclosed provokingly little—was singing to herself, as she stooped over an open hive, and lifted the crowded brood-frames one by one up into the light of day.

"The great work of the year is just beginning with us," explained the bee-mistress. "In these first warm days of spring every hive must be opened and its condition ascertained. Those that are short of stores must be fed; backward colonies must be quickened to a sense of their responsibilities. Clean hives must be substituted for the old, winter-soiled dwellings. Queens that are past their prime will have to be dethroned, and their places filled by younger and more vigorous successors. But it is all typically women's work. You have an old acquaintance with the lordly bee-master and his ways; now come and see how a woman manages."

We passed over to the singing lady in the veil, and—from a safe distance—watched her at her work. Each frame, as it was raised out of the seething abyss of the hive, was turned upside down and carefully examined. A little vortex of bees swung round her head, shrilling vindictively. Those on the uplifted comb-frames hustled to and fro like frightened sheep, or crammed themselves head foremost into the empty cells, out of reach of the disturbing light.

" That is a queenless stock," said the bee-mistress. " It is going to be united with another colony, where there is a young, high-mettled ruler in want of subjects."

We watched the bee-gardener as she went to one of the neighbouring hives, subdued and opened it, drew out all the brood-combs, and brought them over in a carrying-rack, with the bees clustering in thousands all about them. Then a scent-diffuser was brought into play, and the fragrance of lavender-water came over to us, as the combs of both hives were quickly sprayed with the perfume, then lowered into the hive, a frame from each stock alternately. It was the old time-honoured plan for uniting bee-colonies, by impregnating them with the same odour, and so inducing the bees to live together peaceably, where otherwise a deadly war might ensue. But the whole operation was carried through with a neat celerity, and light, dexterous handling, I had never seen equalled by any man.

" That girl," said the bee-mistress, as we moved away, " came to me out of a London office a year ago, anæmic, pale as the paper she typed on all day for a living. Now she is well and strong, and almost as brown as the bees she works among so willingly. All my girls here have come to me from time to time in the same way out of the towns, forsaking indoor employment that was surely stunting all growth of mind and body. And there are thousands who would do the same to-morrow, if only the chance could be given them."

We stopped in the centre of the old orchard. Overhead the swelling fruit-buds glistened against the blue sky. Merry thrush-music rang out far and near. Sun and shadow, the song of the bees, laughing voices, a snatch of an old Sussex chantie, the perfume of violet-beds and nodding gillyflowers, all came over to us through the lichened tree-stems, in a flood of delicious colour and scent and sound. The bee-mistress turned to me, triumphantly.

" Would any sane woman," she asked, " stop in the din and dirt of a smoky city, if she could come and work in a place like this? Bee-keeping for women! do you not see what a chance it opens up to poor toiling folk, pining for fresh air and sunshine, especially to the office-girl class, girls often of birth and refinement—just that kind of poor gentlewomen whose breeding and social station render them most difficult of all to help? And here is work for them, clean, intellectual, profitable; work that will keep them all day long in the open air; a healthy, happy country life, humanly within the reach of all."

" What is wanted," continued the bee-mistress, as we went slowly down the broad main-way of the honey-farm, " is for some great lady, rich in business ideas as well as in pocket, to take up the whole scheme, and to start a network of small bee-gardens for women over the whole land. Very large bee-farms are a mistake, I think, except in the most favourable districts. Bees work only within a radius of two or three miles at most, so that the number of hives that can be kept profitably in a given area has its definite limits. But there is still plenty of room everywhere for bee-farms of moderate size, conducted on the right principles; and there is no reason at all why they should not work together on the co-operative plan, sending all their produce to some convenient centre in each district, to be prepared and marketed for the common good."

" But the whole outcome," she went on, " of a scheme like this depends on the business qualities imported into it. Here, in the heart of the Sussex Weald, we labour together in the midst of almost ideal surroundings, but we never lose sight of the plain, commercial aspect of the thing. We study all the latest writings on our subject, experiment with all novelties, and keep ourselves well abreast of the times in every way. Our system is to make each hive show a clear, definite profit. The annual income is not, and can never be, a very large one, but we fare quite simply, and have sufficient for our needs. In any case, however, we have proved here that a few women, renting a small house and garden out in the country, can live together comfortably on the proceeds from their bees; and there is no reason in the world why the idea should not be carried out by others with equal success."

We had made the round of the whole busy,

murmuring enclosure, and had come again to the little door in the wall. Passing through and out once more into the world of merely masculine endeavour, the bee-mistress gave me a final word.

" You may think," said she, " that what I advocate, though successful in our own single instance, might prove impracticable on a widely extended scale. Well, do you know that last year close upon three hundred and fifty tons of honey were imported into Great Britain from foreign sources,* just because our home apiculturists were unable to cope with the national demand? And this being so, is it too much to think that, if women would only band themselves together and take up bee-keeping systematically, as we have done, all or most of that honey could be produced—of infinitely better quality—here, on our own British soil? "

* Before the War.

CHAPTER V

A BEE-MAN OF THE 'FORTIES

THE old bee-garden lay on the verge of the wood. Seen from a distance it looked like a great white china bowl brimming over with roses; but a nearer view changed the porcelain to a snowy barrier of hawthorn, and the roses became blossoming apple-boughs, stretching up into the May sunshine, where all the bees in the world seemed to have for-gathered, filling the air with their rich wild chant.

Coming into the old garden from the glare of the dusty road, the hives themselves were the last thing to rivet attention. As you went up the shady moss-grown path, perhaps the first impression you became gratefully conscious of was the slow dim quiet of the place—a quiet that had in it all the essentials of silence, and yet was really made up of a myriad blended sounds. Then the sheer carmine of the tulips, in the sunny vista beyond the orchard, came upon you like a trumpet-note through the shadowy aisles of the trees; and after this, in turn, the flaming amber of the marigolds, broad zones of forget-me-nots like strips of the blue sky fallen, snow-drifts of arabis and starwort, purple

44

pansy-spangles veering to every breeze. And last of all you became gradually aware that every bright nook or shade-dappled corner round you had its nestling bee-skep, half hidden in the general riot of blossom, yet marked by the steadier, deeper song of the homing bees.

To stand here, in the midst of the hives, of a fine May morning, side by side with the old bee-man, and watch with him for the earliest swarms of the year, was an experience that took one back far into another and a kindlier century. There were certain hives in the garden, grey with age and smothered in moss and lichen, that were the traditional mother-colonies of all the rest. The old bee-keeper treasured them as relics of his sturdy manhood, just as he did the percussion fowling-piece over his mantel; and pointed to one in particular as being close on thirty years old. Nowadays remorseless science has proved that the individual life of the honey-bee extends to four or five months at most; but the old bee-keeper firmly believed that some at least of the original members of this colony still flourished in green old age deep in the sombre corridors of the ancient skep. Bending down, he would point out to you, among the crowd on the alighting-board, certain bees with polished thorax and ragged wings worn almost to a stump. While the young worker-bees were charging in and out of the hive at breakneck speed, these superannuated amazons doddered about in the sunlight, with an obvious and pathetic assumption of importance. They were really the last survivors of the bygone winter's brood. Their task of hatching the new spring generation was over; and now, the power of flight denied to them, they busied themselves in the work of sentinels at the gate, or in grooming the young bees as they came out for their first adventure into the far world of blossoming clover under the hill.

For modern apiculture, with its interchangeable comb-frames and section-supers, and American notions generally, the old bee-keeper harboured a fine contempt. In its place he had an exhaustless store of original bee-knowledge, gathered throughout his sixty odd years of placid life among the bees. His were all old-fashioned hives of straw, hackled and potsherded just as they must have been any time since Saxon Alfred burned the cakes. Each bee-colony had its separate three-legged stool, and each leg stood in an earthen pan of water, impassable moat for ants and " wood-li's," and such small honey-thieves. Why the hives were thus dotted about in such admired but inconvenient disorder was a puzzle at first, until you learned more of ancient bee-traditions. Wherever a swarm settled—up in the pink-rosetted apple-boughs, under the eaves of the old thatched cottage, or deep in the tangle of the hawthorn hedge—there, on the nearest open ground beneath, was its inalienable, predetermined home. When, as sometimes happened, the swarm went straight away out of sight over the meadows, or sailed off like a pirouetting grey cloud over the roof of the wood, the old bee-keeper never sought to reclaim it for the garden.

" 'Tis gone to the shires fer change o' air," he would say, shielding his bleak blue eyes with his hand, as he gazed after it. " 'Twould be agen natur' to hike 'em back here along. An' naught but ill-luck an' worry wi'out end.''

He never observed the skies for tokens of to-morrow's weather, as did his neighbours of the countryside. The bees were his weather-glass and thermometer in one. If they hived very early after noon, though the sun went down in clear gold and the summer night loomed like molten amethyst under the starshine, he would prophesy rain before morning. And sure enough you were wakened at dawn by a furious patter on the window, and the booming of the south-west wind in the pine-clad crest of the hill. But if the bees loitered afield far into the gusty crimson gloaming, and the loud darkness that followed seemed only to bring added intensity to the busy labour-note within the hives, no matter how the wind keened or the griddle of black storm-cloud threatened, he would go on with his evening task of watering his garden, sure of a morrow of cloudless heat to come.

He knew all the sources of honey for miles around; and, by taste and smell, could decide at once the particular crop from which each sample had been gathered. He would discriminate between that from white clover or sainfoin; the produce of the yellow charlock wastes; or the orchard-honey, wherein it seemed the fragrance of cherry-bloom was always to be differentiated from that of apple or damson or pear. He would tell you when good honey had been spoilt by the grosser flavour of sunflower or horse-chestnut; or when the detestable honey-dew had entered into its composition; or, the super-caps having been removed too late in the season, the bees had got at the early ivy-blossom, and so degraded all the batch.

Watching bees at work of a fair morning in May, nothing excites the wonder of the casual looker-on more than the mysterious burdens they are for ever bringing home upon their thighs; semi-globular packs, always gaily coloured, and often so heavy and cumbersome that the bee can hardly drag its weary way into the hive. This is pollen, to be stored in the cells, and afterwards kneaded up with honey as food for the young bees. The old man could say at once by the colour from which flower each load was obtained. The deep brown-gold panniers came from the gorse-bloom; the pure snow-white from the hawthorns; the vivid yellow, always so big and seemingly so weighty, had been filled in the buttercup meads. Now and again, in early spring, a bee would come blundering home with a load of pallid sea-green hue. This came from the gooseberry bushes. And later, in summer, when the poppies began to throw their scarlet shuttles in the corn, many of these airy cargoes would be of a rich velvety black. But there was one kind which the old bee-man had never yet succeeded in tracing to its flowery origin. He saw it only rarely, perhaps not a dozen times in the season—a wonderful deep rose-crimson, singling out its bearer, on her passage through the throng, as with twin danger-lamps, doubly bright in the morning glow.

Keeping watch over the comings and goings of his bees was always his favourite pastime, year in and year out; but it was in the later weeks of May that his interest in them culminated. He had always had swarms in May as far back as his memory could serve him; and the oldest hive in the garden was generally the first to swarm. As a rule the bees gave sufficient warning of their intended migration some hours before their actual issue. The strenuous pell-mell business of the hive would come to a sudden portentous halt. While a few of the bees still darted straight off into the sunshine on their wonted errands, or returned with the usual motley loads upon their thighs, the rest of the colony seemed to have abandoned work altogether. From early morning they hung in a great brown cluster all over the face of the hive, and down almost to the earth beneath; a churning mass of insect-life that grew bigger and bigger with every moment, glistening like wet seaweed in the morning sun. In the cluster itself there was an uncanny silence. But out of the depths of the hive came a low vibrating murmur, wholly distinct from its usual note; and every now and again a faint shrill piping sound could be heard, as the old queen worked herself up to swarming frenzy, vainly seeking the while to reach the royal nursery where the rival who was to oust her from her old dominion was even then steadily gnawing through her constraining prison walls.

At these momentous times a quaint ceremonial was rigidly adhered to by the old bee-master. First he brought out a pitcher of home-brewed ale, from which all who were to assist in the swarm-taking were required to drink, as at a solemn rite. The dressing of the skep was his next care. A little of the beer was sprinkled over its interior, and then it was carefully scoured out with a handful of balm and lavender and mint. After this the skep was covered up and set aside in the shade; and the old bee-keeper, carrying an ancient battered copper bowl in one gnarled hand, and a great door-key in the other, would lead the way towards the hive, his drab smock-frock mowing the scarlet tulip-heads down as he went.

Sometimes the swarm went off without any preliminary warning, just as if the skep had burst like a bombshell, volleying its living contents into the sky. But oftener it went through the several stages of a regular process. After much waiting and many false alarms, a peculiar stir would come in the throng of bees cumbering the entrance to the hive. Thousands rose on the wing, until the sunshine overhead was charged with them as with countless fluttering atoms of silver-foil; and a wild joyous song spread far and wide, overpowering all other sounds in the garden. Within the hive the rich bass note had ceased; and a hissing noise, like a great caldron boiling over, took its place, as the bees inside came pouring out to join the carolling multitude above. Last of all came the queen. Watching for her through the glittering gauzy atmosphere of flashing wings, she was always strangely conspicuous, with her long pointed body of brilliant chestnut-red. She came hustling forth; stopped for an instant to comb her antennæ on the edge of the foot-board; then soared straight up into the blue, the whole swarm crowding deliriously in her train.

Immediately the old bee-man commenced a weird tom-tomming on his metal bowl. " Ringing the bees " was an exact science with him. They were supposed to fly higher or lower according to the measure of the music; and now the great door-key beat out a slow, stately chime like a cathedral bell. Whether this ringing of the old-time skeppists had any real influence on the movements of a swarm has never been absolutely determined; but there was no doubt in this case of the beè-keeper's perfect faith in the process, or that the bees would commence their descent and settle, usually in one of the apple trees, very soon after the din began.

The rapid growth of the swarm-cluster was always one of the most bewildering things to watch. From a little dark knot no bigger than the clenched hand, it swelled in a moment to the size of a half-gallon measure, growing in girth and length with inconceivable swiftness, until the branch began to droop under its weight. A minute more, and the last of the flying bees had joined the cluster; the stout apple-branch was bent almost double; and the completed swarm hung within a few inches of the ground, a long cigar-shaped mass gently swaying to and fro in the flickering light and shade.

The joyous trek-song of the bees, and the clanging melody of key and basin, died down together. The old murmuring, songful quiet closed over the garden again, as water over a cast stone. To hive a swarm thus easily within reach was a simple matter. Soon the old bee-man had got all snugly inside the skep, and the hive in its self-appointed station. And already the bees were settling down to work; hovering merrily about it, or packed in the fragrant darkness busy at comb-building, or lancing off to the clover-fields, eager to begin the task of provisioning the new home.

CHAPTER VI

HEREDITY IN THE BEE-GARDEN

WE were in the great high-road of Warrilow bee-farm, and had stopped midway down in the heart of the waxen city. On every hand the hives stretched away in long trim rows, and the hot June sunshine was alive with darting bees and fragrant with the smell of new-made honey.

" Swarming? " said the bee-master, in answer to a question I had put to him. " We never allow swarming here. My bees have to work for me, and not for themselves; so we have discarded that old-fashioned notion long ago."

He brought his honey-barrow to a halt, and sat down ruminatively on the handle.

" Swarming," he went on to explain, " is the great trouble in modern bee-keeping. It is a bad legacy left us by the old-time skeppists. With the ancient straw hives and the old benighted methods of working, it was all very well. When bee-burning was the custom, and all the heaviest hives were foredoomed to the sulphur-pit, the best bees were those that gave the earliest and the largest swarms. The more stocks there were in the garden the more honey there would be for market.

Swarming was encouraged in every possible way.
And so, at last, the steady, stay-at-home variety of
honey-bee became exterminated, and only the
inveterate swarmers were kept to carry on the
strain."

I quoted the time-honoured maxim about a
swarm in May being worth a load of hay. The
bee-master laughed derisively.

" To the modern bee-keeper," he said, " a
swarm in May is little short of a disgrace. There
is no clearer sign of bad beemanship nowadays than
when a strong colony is allowed to weaken itself
by swarming on the eve of the great honey-flow,
just when strength and numbers are most needed.
Of course, in the old days, the maxim held true
enough. The straw skeps had room only for a
certain number of bees, and when they became too
crowded there was nothing for it but to let the
colonies split up in the natural way. But the
modern frame-hive, with its extending brood-
chamber, does away with that necessity. Instead
of the old beggarly ten or twelve thousand, we can
now raise a population of forty or fifty thousand
bees in each hive, and so treble and quadruple the
honey-harvest."

" But," I asked him, " do not the bees go on
swarming all the same, if you let them? "

" The old instincts die hard," he said. " Some
day they will learn more scientific ways; but as
yet they have not realised the change that modern
bee-keeping has made in their condition. Of
course, swarming has its clear, definite purpose,
apart from that of relieving the congestion of the
stock. When a hive swarms, the old queen goes

off with the flying squadron, and a new one takes her place at home. In this way there is always a young and vigorous queen at the head of affairs, and the well-being of the parent stock is assured. But advanced bee-keepers, whose sole object is to get a large honey yield, have long recognised that this is a very expensive way of rejuvenating old colonies. The parent hive will give no surplus honey for that season; and the swarm, unless it is a large and very early one, will do little else than furnish its brood-nest for the coming winter. But if swarming be prevented, and the stock requeened artificially every two years, we keep an immense population always ready for the great honey-flow, whenever it begins."

He took up the heavy barrow, laden with its pile of super-racks, and started trundling it up the path, talking as he went.

"If only the bees could be persuaded to leave the queen-raising to the bee-keeper, and would attend to nothing else but the great business of honey-getting! But they won't—at least, not yet. Perhaps in another hundred years or so the old wild habits may be bred out of them; but at present it is doubtful whether they are conscious of any 'keeping' at all. They go the old tried paths determinedly; and the most that we can accomplish is to undo that part of their work which is not to our liking, or to make a smoother road for them in the direction they themselves have chosen."

"But you said just now," I objected, "that no swarming was allowed among your bees. How do you manage to prevent it?"

"It is not so much a question of prevention as of

cure. Each hive must be watched carefully from the beginning. From the time the queen commences to lay, in the first mild days of spring, we keep the size of the brood nest just a little ahead of her requirements. Every week or two I put in a new frame of empty combs, and when she has ten frames to work upon, and honey is getting plentiful, I begin to put on the store-racks above, just as I am doing now. This will generally keep them to business; but with all the care in the world the swarming fever will sometimes set in. And then I always treat it in this way."

He had stopped before one of the hives, where the bees were hanging in a glistening brown cluster from the alighting-board; idling while their fellows in the bee-garden seemed all possessed with a perfect fury of work. I watched him as he lighted the smoker, a sort of bellows with a wide tin funnel packed with chips of dry rotten wood. He stooped over the hive, and sent three or four dense puffs of smoke into the entrance.

" That is called subduing the bees," he explained, " but it really does nothing of the kind. It only alarms them, and a frightened bee always rushes and fills herself with honey, to be ready for any emergency. She can imbibe enough to keep her for three or four days; and once secure of immediate want, she waits with a sort of fatalistic calm for the development of the trouble threatening."

He halted a moment or two for this process to complete itself, then began to open the hive. First the roof came off; then the woollen quilts and square of linen beneath were gradually peeled from the tops of the comb-frames, laying bare the interior of the hive. Out of its dim depths came up a steady rumbling note like a train in a tunnel, but only a few of the bees got on the wing and began to circle round our heads viciously. The frames hung side by side, with a space of half an inch or so between. The bee-master lifted them out carefully one by one.

" Now, see here," he said, as he held up the first frame in the sunlight, with the bees clinging in thousands to it, " this end comb ought to have nothing but honey in it, but you see its centre is covered with brood-cells. The queen has caught the bee-man napping, and has extended her nursery to the utmost limit of the hive. She is at the end of her tether, and has therefore decided to swarm. Directly the bees see this they begin to prepare for the coming loss of their queen by raising another, and to make sure of getting one they always breed three or four."

He took out the next comb and pointed to a round construction, about the size and shape of an acorn, hanging from its lower edge.

" That is a queen cell; and here, on the next comb, are two more. One is sealed over, you see, and may hatch out at any moment; and the others are nearly ready for closing. They are always carefully guarded, or the old queen would destroy them. And now to put an end to the swarming fit."

He took out all the combs but the four centre ones; and, with a goose wing, gently brushed the bees off them into the hive. The six combs were then taken to the extricating-house hard by. The sealed honey-cells on all of them were swiftly uncapped, and the honey thrown out by a turn or two in the centrifugal machine. Now we went back to the hive. Right in the centre the bee-master put a new, perfectly empty comb, and on each side of this came the four principal brood frames with the queen still on them. Outside of these again the combs from which we had extracted all the honey were brought into position. And then a rack of new sections was placed over all, and the hive quickly closed up. The entire process seemed the work of only a few minutes.

" Now," said the bee-master triumphantly, as he took up his barrow again, "we have changed the whole aspect of affairs. The population of the hive is as big as ever; but instead of a house of plenty it is a house of dearth. The larder is empty, and the only cure for impending famine is hard work; and the bees will soon find that out and set to again. Moreover, the queen has now plenty of room for laying everywhere, and those exasperating prison-cradles, with her future rivals hatching in them, have been done away with. She has no further reason for flight, and the bees, having had all their preparations destroyed, have the best of reasons for keeping her. Above all, there is the new super-rack, greatly increasing the hive space, and they will be given a second and third rack, or even a fourth one, long before they feel the want of it. Every motive for swarming has been removed, and the result to the bee-master will probably be seventy or eighty pounds of surplus honey, instead of none at all, if the bees had been left to their old primæval ways."

" You must always remember, however," he added, as a final word, " that bees do nothing invariably. 'Tis an old and threadbare saying amongst bee-keepers, but there's nothing truer under the sun. Bees have exceptions to almost every rule. While all other creatures seem to keep blindly to one pre-ordained way in everything they do, you can never be certain at any time that bees will not reverse their ordinary course to meet circumstances you may know nothing of. And that is all the more reason why the bee-master himself should allow no deviations in his own work about the hives: his ways must be as the ways of the Medes and Persians.''

CHAPTER VII

NIGHT ON A HONEY-FARM

THE sweet summer dusk was over the bee-farm. On every side, as I passed through, the starlight showed me the crowding roofs of the city of hives; and beyond these I could just make out the dim outline of the extracting-house, with a cheerful glow of lamplight streaming out from window and door. The rumble of machinery and the voices of the bee-master and his men grew louder as I approached. A great business seemed to be going forward within. In the centre of the building stood a strange-looking engine, like a brewer's vat on legs. It was eight or nine feet broad and some five feet high; and a big horizontal wheel lay within the great circle, completely filling its whole circumference. As I entered, the wheel was going round with a deep reverberating noise as fast as two strong men could work the gearing; and the bee-master stood close by, carefully timing the operation.

"Halt!" he shouted. The great wheel-of-fortune stopped. A long iron bar was pulled down and the wheel rose out of the vat. Now I could see that its whole outer periphery was covered with frames of honeycomb, each in its separate gauze-wire cage. The bee-master tugged a lever. The cages—there must have been twenty-five or thirty of them—turned over simultaneously like single leaves of a book, bringing the other side of each comb into place. The wheel dropped down once more, and swung round again on its giddy journey. From my place by the door I could hear the honey driving out against the sides of the vat like heavy rain.

"Halt!" cried the bee-master again. Once more the big wheel rose, glistening and dripping, into the yellow lamplight. And now a trolley was pushed up laden with more honeycomb ready for extraction. The wire-net cages were opened, the empty combs taken out, and full ones deftly put in their place. The wheel plunged down again into its mellifluous cavern, and began its deep song once more. The bee-master gave up his post to the foreman, and came towards me, wiping the honey from his hands. He was very proud of his big extractor, and quite willing to explain the whole process. "In the old days," he said, "the only way to get the honey from the comb was to press it out. You could not obtain your honey without destroying the comb, which at this season of the year is worth very much more than the honey itself; for if the combs can be emptied and restored perfect to the hive, the bees will fill them again immediately, without having to waste valuable time in the height of the honey-flow by stopping to make new combs. And when the bees are wax-making they are not only prevented from gathering honey, but have to consume their own stores. While they are making one pound of comb they will eat seventeen or eighteen pounds of honey. So the man who hit upon the idea of drawing the honey from the comb by centrifugal force did a splendid thing for modern bee-farming. English honey was nothing until the extractor came and changed bee-keeping from a mere hobby into an important industry. But come and see how the thing is done from the beginning."

He led the way towards one end of the building. Here three or four men were at work at a long table surrounded by great stacks of honeycombs in their oblong wooden frames. The bee-master took up one of these. " This," he explained, " is the bar-frame just as it comes from the hive. Ten of them side by side exactly fill a box that goes over the hive proper. The queen stays below in the brood-nest, but the worker bees come to the top to store the honey. Then, every two or three days, when the honey-flow is at its fullest, we open the super, take out the sealed combs, and put in combs that have been emptied by the extractor. In a few days these also are filled and capped by the bees, and are replaced by more empty combs in the same way; and so it goes on to the end of the honey-harvest."

We stood for a minute or two watching the work at the table. It went on at an extraordinary pace. Each workman seized one of the frames and poised it vertically over a shallow metal tray. Then, from a vessel of steaming hot water that stood at his elbow, he drew the long, flat-headed Bingham knife, and with one swift slithering cut removed the whole of the cell-cappings from the surface of the comb. At once the knife was thrown back into its smoking bath, and a second one taken out, with which the other side of the comb was treated. Then the comb was hung in the rack of the trolley, and the keen hot blades went to work on another frame. As each trolley was fully loaded it was whisked off to the extracting-machine and another took its place.

" All this work," explained the bee-master, as we passed on, " is done after dark, because in the daytime the bees would smell the honey and would besiege us. So we cannot begin extracting until they are all safely hived for the night." He stopped before a row of bulky cylinders. " These," he said, " are the honey ripeners. Each of them holds about twenty gallons, and all the honey is kept here for three or four days to mature before it is ready for market. If we were to send it out at once it would ferment and spoil. In the top of each drum there are fine wire strainers, and the honey must run through these, and finally through thick flannel, before it gets into the cylinder. Then, when it is ripe, it is drawn off and bottled."

One of the big cylinders was being tapped at the moment. A workman came up with a kind of gardener's water-tank on wheels. The valve of the honey-vat was opened, and the rich fluid came gushing out like liquid amber. " This is all white-clover honey," said the bee-master, tasting it critically. " The next vat there ought to be pure sainfoin. Sometimes the honey has a distinct almond flavour; that is when hawthorn is abundant. Honey varies as much as wine. It is good or bad according to the soil and the season. Where the horse-chestnut is plentiful the honey has generally a rank taste. But this is a sheep-farmers' country, where they grow thousands of acres of rape and lucerne and clover for sheep-feed; and nothing could be better for the bees."

By this time the gardener's barrow was full to the brim. We followed it as it was trundled heavily away to another part of the building. Here a little company of women were busy filling the neat glass jars, with their bright screw-covers of tin; pasting on the label of the big London stores, whither most of the honey was sent; and packing the jars into their travelling-cases ready for the railway-van in the morning. The whole place reeked with the smell of new honey and the faint, indescribable odour of the hives. As we passed out of the busy scene of the extracting-house into the moist dark night again, this peculiar fragrance struck upon us overpoweringly. The slow wind was setting our way, and the pungent odour from the hives came up on it with a solid, almost stifling, effect.

"They are fanning hard to-night," said the bee-master, as we stopped halfway down the garden. "Listen to the noise they're making!"

The moon was just tilting over the tree-tops. In its dim light the place looked double its actual size. We seemed to stand in the midst of a great town of bee-dwellings, stretching vaguely away into the darkness. And from every hive there rose the clear deep murmur of the ventilating bees.

The bee-master lighted his lantern, and held it down close to the entrance of the nearest hive.

"Look how they form up in rows, one behind the other, with their heads to the hive; and all fanning with their wings! They are drawing the hot air out. Inside there is another regiment of them, but those are facing the opposite way, and drawing the cool air in. And so they keep the hive always at the right temperature for honey-making, and for hatching out the young bees."

"Who was it," he asked ruminatively, as the gate of the bee-farm closed at last behind us, and we were walking homeward through the glimmering dusk of the lane—" who was it first spoke of the ' busy bee '? Busy! 'Tis not the word for it! Why, from the moment she is born to the day she dies the bee never rests nor sleeps! It is hard work night and day, from the cradle-cell to the grave; and in the honey-season she dies of it after a month or so. It is only the drone that rests. He is very like some humans I know of his own sex; he lives an idle life, and leaves the work to the womenkind. But the drone has to pay for it in the end, for the drudging woman-bee revolts sooner or later. And then she kills him. In bee-life the drone always dies a violent death; but in human life—well, it seems to me a little bee-justice wouldn't be amiss with some of them."

CHAPTER VIII

IN A BEE-CAMP

"'TIS a good thing—life; but ye never know how good, really, till you've followed the bees to the heather."

It was an old saying of the bee-master's, and it came again slowly from his lips now, as he knelt by the camp-fire, watching the caress of the flames round the bubbling pot. We were in the heart of the Sussex moorland, miles away from the nearest village, still farther from the great bee-farm where, at other times, the old man drove his thriving trade. But the bees were here—a million of them perhaps—all singing their loudest in the blossoming heather that stretched away on every side to the far horizon, under the sweltering August sun.

Getting the bees to the moors was always the chief event of the year down at the honey-farm. For days the waggons stood by the laneside, all ready to be loaded up with the best and most populous hives; but the exact moment of departure depended on one very uncertain factor. The white-clover crop was almost at an end. Every day saw the acreage of sainfoin narrowing, as the sheep-

65

folds closed in upon it, leaving nothing but bare yellow waste, where had been a rolling sea of crimson blossom. But the charlock lay on every hillside like cloth-of-gold. Until harvest was done the fallows were safe from the ploughshare, and what proved little else than a troublesome weed to the farmer was like golden guineas growing to every keeper of bees.

But at last the new moon brought a sharp chilly night with it, and the long-awaited signal was given. Coming down with the first grey glint of morning from the little room under the thatch, I found the bee-garden in a swither of commotion. A faint smell of carbolic was on the air, and the shadowy figures of the bee-master and his men were hurrying from hive to hive, taking off the super-racks that stood on many three and four stories high. The honey-barrows went to and fro groaning under their burdens; and the earliest bees, roused from their rest by this unwonted turmoil, filled the grey dusk with their high timorous note.

The bee-master came over to me in his white overalls, a weird apparition in the half-darkness. " 'Tis the honey-dew," he said, out of breath, as he passed by. "The first cold night of summer brings it out thick on every oak-leaf for miles around; and if we don't get the supers off before the bees can gather it, the honey will be blackened and spoiled for market."

He carried a curious bundle with him, an armful of fluttering pieces of calico, and I followed him as he went to work on a fresh row of hives. From each bee-dwelling the roof was thrown off, the inner coverings removed, and one of the squares of cloth—damped with the carbolic solution—quickly drawn over the topmost rack. A sudden fearsome buzzing uprose within, and then a sudden silence. There is nothing in the world a bee dreads more than the smell of carbolic acid. In a few seconds the super-racks were deserted, the bees crowding down into the lowest depths of the hives. The creaking barrows went down the long row in the track of the master, taking up the heavy racks as they passed. Before the sun was well up over the hill-brow the last load had been safely gathered in, and the chosen hives were being piled into the waggons, ready for the long day's journey to the moors.

All this was but a week ago; yet it might have been a week of years, so completely had these rose-red highland solitudes accepted our invasion, and absorbed us into their daily round of sun and song. Here, in a green hollow of velvet turf, right in the heart of the wilderness, the camp had been pitched—the white bell-tents with their skirts drawn up, showing the spindle-legged field-bedsteads within; the filling-house, made of lath and gauze, where the racks could be emptied and recharged with the little white wood section-boxes, safe from marauding bees; the honey-store, with its bee-proof crates steadily mounting one upon the other, laden with rich brown heather-honey—the finest sweet-food in the world. And round the camp, in a vast spreading circle, stood the hives—a hundred or more—knee-deep in the rosy thicket, each facing outward, and each a whirling vortex of life from early dawn to the last amber gleam of sunset abiding under the flinching silver of the stars.

The camp-fire crackled and hissed, and the pot
sent forth a savoury steam into the morning air.
From the heather the deep chant of busy thousands
came over on the wings of the breeze, bringing
with it the very spirit of serene content. The bee-
master rose and stirred the pot ruminatively.

" B'iled rabbit!" said he, looking up, with the
light of old memories coming in his gnarled brown
face. " And forty years ago, when I first came to
the heather, it used to be b'iled rabbit too. We
could set a snare in those days as well as now.
But 'twas only a few hives then, a dozen or so of
old straw skeps on a barrow, and naught but the
starry night for a roof-tree, or a sack or two to
keep off the rain. None of your women's luxuries
in those times!"

He looked round rather disparagingly at his own
tent, with its plain truckle-bed, and tin wash-bowl,
and other deplorable signs of effeminate self-
indulgence.

" But there was one thing," he went on, " one
thing we used to bring to the moors that never
comes now. And that was the basket of sulphur-
rag. When the honey-flow is done, and the waggons
come to fetch us home again, all the hives will go
back to their places in the garden none the worse
for their trip. But in the old days of bee-burning
never a bee of all the lot returned from the moors.
Come a little way into the long grass yonder, and
I'll show ye the way of it."

With a stick he threshed about in the dry bents,
and soon lay bare a row of circular cavities in the
ground. They were almost choked up with moss
and the rank undergrowth of many years; but

originally they must have been each about ten inches broad by as many deep.

" These," said the bee-master, with a shamefaced air of confession, " were the sulphur-pits. I dug them the first year I ever brought hives to the heather; and here, for twenty seasons or more, some of the finest and strongest stocks in Sussex were regularly done to death. 'Tis a drab tale to tell, but we knew no better then. To get the honey away from the bees looked well-nigh impossible with thousands of them clinging all over the combs. And it never occurred to any of us to try the other way, and get the bees to leave the honey. Yet bee-driving, 'tis the simplest thing in the world, as every village lad knows to-day."

We strolled out amongst the hives, and the bee-master began his leisurely morning round of inspection. In the bee-camp, life and work alike took their time from the slow march of the summer sun, deliberate, imperturbable, across the pathless heaven. The bees alone keep up the heat and burden of the day. While they were charging in and out of the hives, possessed with a perfect fury of labour, the long hours of sunshine went by for us in immemorial calm. Like the steady rise and fall of a windless tide, darkness and day succeeded one another; and the morning splash in the dew-pond on the top of the hill, and the song by the camp-fire at night, seemed divided only by a dim formless span too uneventful and happy to be called by the old portentous name of Time.

And yet every moment had its business, not to be delayed beyond its imminent season. Down in the bee-farm the work of honey-harvesting always carried with it a certain stress and bustle. The great centrifugal extractor would be roaring half the night through, emptying the super-combs, which were to be put back into the hives on the morrow, and refilled by the bees. But here, on the moors, modern bee-science is powerless to hurry the work of the sunshine. The thick heather-honey defies the extracting-machine, and cannot be separated without destroying the comb. Moorland honey— except where the wild sage is plentiful enough to thin down the heather sweets—must be left in the virgin comb; and the bee-man can do little more than look on as vigilantly as may be at the work of his singing battalions, and keep the storage-space of the hives always well in advance of their need.

Yet there is one danger—contingent at all seasons of bee-life, but doubly to be guarded against during the critical time of the honey-flow.

As we loitered round the great circle, the old bee-keeper halted in the rear of every hive to watch the contending streams of workers, the one rippling out into the blue air and sunshine, the other setting more steadily homeward, each bee weighed down with her load of nectar and pale grey pollen, as she scrambled desperately through the opposing crowd and vanished into the seething darkness within. As we passed each hive, the old bee-man carefully noted its strength and spirit, comparing it with the condition of its neighbours on either hand. At last he stopped by one of the largest hives, and pointed to it significantly.

" Can ye see aught amiss? " he asked, hastily rolling his shirt-sleeves up to the armpit.

I looked, but could detect nothing wrong. The

multitude round the entrance to this hive seemed
larger and busier than with any other, and the note
within as deeply resonant.

" Ay! they're erpulous enough," said the bee-
master, as he lighted his tin-nozzled bellows-smoker
and coaxed it into full blast. " But hark to the din!
'Tis not work this time; 'tis mortal fear of something. Flying strong? Ah, but only a yard or
two up, and back again. There's trouble at hand,
and they've only just found it out. The matter is,
they have lost their queen."

He was hurriedly removing the different parts of
the hive as he spoke. A few quick puffs from the
smoker were all that was needed at such a time.
With no thought but for the tragedy that had come
upon them, the bees were rushing madly to and fro
in the hive, not paying the slightest attention to the
fact that their house was falling asunder piecemeal
and the sudden sunshine riddling it through and
through, where had been nothing but Cimmerian
darkness before. Under the steady slow hand of
the master, the teeming section-racks came off one
by one, until the lowest chamber—the nursery of
the hive—was reached, and a note like imprisoned
thunder in miniature burst out upon us.

The old bee-keeper lifted out the brood-frames,
and subjected each to a lynx-eyed scrutiny. At last
he dived his bare hand down into the thick of the
bees, and brought up something to show me. It
was the dead queen; twice the size of all the rest,
with short oval wings and a shining red-gold body,
strangely conspicuous among the score or so of
dun-coloured workers which still crowded round her
on the palm of his hand.

" In the old days," said the bee-master, " before the movable-comb hive was invented, if the queen died like this, it would throw the whole colony out of gear for the rest of the season. Three weeks must elapse before a new queen could be hatched and got ready for work; and then the honey-harvest would be over. But see how precious time can be saved under the modern system."

He led the way to a hive which stood some distance apart from the rest. It was much smaller than the others, and consisted merely of a row of little boxes, each with its separate entrance, but all under one common roof. The old bee-man opened one of the compartments, and lifted out its single comb-frame, on which were clustered only a few hundred bees. Searching among these with a wary forefinger, at last he seized one by the wings and held it up to view.

" This is a spare queen," said he. " 'Tis always wise to bring a few to the heather, against any mischance. And now we'll give her to the motherless bees; and in an hour or two the stock will be at work again as busily as ever."

CHAPTER IX

THE BEE-HUNTERS

"IN that bit of forest," said the bee-master, indicating a long stretch of neighbouring woodland with one comprehensive sweep of his thumb, " there are tons of honey waiting for any man who knows how to find it."

I had met and stopped the old bee-keeper and his men, bent on what seemed a rather singular undertaking. They carried none of the usual implements of their craft, but were laden up with the paraphernalia of woodmen—rip-saws and hatchets and climbing-irons, and a mysterious box or two, the use of which I could not even guess at. But the bee-master soon made his errand plain.

"Tons of honey," he went on. " And we are going to look for some of it. There have been wild bees, I suppose, in the forest country from the beginning of things. Then see how the land lies. There are villages all round, and for ages past swarms have continually got away from the bee-gardens, and hived themselves in the hollow trunks

73

of the trees. Then every year these stray colonies have sent out their own swarms again, until to-day the woods are full of bees, wild as wolves and often as savage, guarding stores that have been accumulating perhaps for years and years."

He shifted his heavy kit from one shoulder to the other. Overhead the sun burned in a cloudless August sky, and the willow-herb by the roadside was full of singing bees and the flicker of white butterflies. In the hedgerows there were more bees plundering the blackberry blossom, or sounding their vagrant note in the white convolvulus-bells which hung in bridal wreaths at every turn of the way. Beyond the hedgerow the yellow cornlands flowed away over hill and dale under the torrid light; and each scarlet poppy that hid in the rustling gold-brown wheat had its winged musician chanting at its portal. As I turned and went along with the expedition, the bee-master gave me more details of the coming enterprise.

" Mind you," he said, " this is not good beemanship as the moderns understand it. It is nothing but bee-murder, of the old-fashioned kind. But even if the bees could be easily taken alive, we should not want them in the apiary. Blood counts in bee-life, as in everything else; and these forest-bees have been too long under the old natural conditions to be of any use among the domestic strain. However, the honey is worth the getting, and if we can land only one big stock or two it will be a profitable day's work."

We had left the hot, dusty lane, and taken to the field-path leading up through a sea of white clover to the woods above.

"This is the after-crop," said the bee-master, as he strode on ahead with his jingling burden. "The second cut of Dutch clover always gives the most honey. Listen to the bees everywhere—it is just like the roar of London heard from the top of St Paul's! And most of it here is going into the woods, more's the pity. Well, well; we must try to get some of it back to-day."

Between the verge of the clover-field and the shadowy depths of the forest ran a broad green waggon-way; and here we came to a halt. In the field we had lately traversed the deep note of the bees had sounded mainly underfoot; but now it was all above us, as the honeymakers sped to and fro between the sunlit plane of blossom and their hidden storehouses in the wood. The upper air was full of their music; but, straining the sight to its utmost, not a bee could be seen.

"And you will never see them," said the bee-master, watching me as he unpacked his kit. "They fly too fast and too high. And if you can't see them go by out here in the broad sunshine, how will you track them to their lair through the dim light under the trees? And yet," he went on, "that is the only way to do it. It is useless to search the wood for their nests; you might travel the whole day through and find nothing. The only plan is to follow the laden bees returning to the hive. And now watch how we do that in Sussex."

From one of the boxes he produced a contrivance like a flat tin saucer mounted on top of a pointed stick. He stuck this in the ground near the edge of the clover-field so that the saucer stood on a level with the highest blossoms. Now he took a small bottle of honey from his pocket, emptied it into the tin receptacle, and beckoned me to come near. Already three or four bees had discovered this unawaited feast and settled on it; a minute more and the saucer was black with crowding bees. Now the bee-master took a wire-gauze cover and softly inverted it over the saucer. Then, plucking his ingenious trap up by the roots, he set off towards the forest with his prisoners, followed by his men.

" These," said he, " are our guides to the secret treasure-chamber. Without them we might look for a week and never find it. But now it is all plain sailing, as you'll see."

He pulled up on the edge of the wood. By this time every bee in the trap had forsaken the honey, and was clambering about in the top of the dome-shaped lid, eager for flight.

" They are all full of honey," said the bee-master, " and the first thing a fully-laden bee thinks of is home. And now we will set the first one on the wing."

He opened a small valve in the trap-cover, and allowed one of the bees to escape. She rose into the air, made a short circle, then sped away into the gloom of the wood. In a moment she was lost to sight, but the main direction of her course was clear; and we all followed helter-skelter until our leader called another halt.

" Now watch this one," he said, pressing the valve again.

This time the guide rose high into the dim air, and was at once lost to my view. But the keen eyes of the old bee-man had challenged her.

"There she goes!" he said, pointing down a long shadowy glade somewhat to his left. "Watch that bit of sunlight away yonder!"

I followed this indication. Through the dense wood-canopy a hundred feet away the sun had thrust one long golden tentacle; and I saw a tiny spark of light flash through into the gloom beyond. We all stampeded after it.

Another and another of the guides was set free, each one taking us deeper into the heart of the forest, until at last the bee-master suddenly stopped and held up his hand.

"Listen!" he said under his breath.

Above the rustling of the leaves, above the quiet stir of the undergrowth and the crooning of the stock-doves, a shrill insistent note came over to us on the gentle wind. The bee-man led the way silently into the darkest depths of the wood. Halting, listening, going swiftly forward in turn, at last he stopped at the foot of an old decayed elm-stump. The shrill note we had heard was much louder now, and right overhead. Following his pointing forefinger, I saw a dark cleft in the old trunk about twenty feet above; and round this a cloud of bees was circling, filling the air with their rich deep labour-song. At the same instant, with a note like the twang of a harp-string, a bee came at me and fastened a red-hot fish-hook into my cheek. The old bee-keeper laughed.

"Get this on as soon as you can," he said, producing a pocketful of bee-veils, and handing me one from the bunch. "These are wild bees, thirty thousand of them, maybe; and we shall need all our armour to-day. Only wait till they find us out! But now rub your hands all over with this."

Every man scrambled into his veil, and anointed his hands with the oil of wintergreen—the one abiding terror of vindictive bees. And then the real business of the day commenced.

The bee-master had strapped on his climbing-irons. Now he struck his way slowly up the tree, tapping the wood with the butt-end of a hatchet inch by inch as he went. At last he found what he wanted. The trunk rang hollow about a dozen feet from the ground. Immediately he began to cut it away. The noise of the hatchet woke all the echoes of the forest. The chips came fluttering to the earth. The rich murmur overhead changed to an angry buzzing. In a moment the bees were on the worker in a vortex of humming fury, covering his veil, his clothes, his hands. But he worked on unconcernedly until he had driven a large hole through the crust of the tree and laid bare the glistening honeycomb within. Now I saw him take from a sling-bag at his side handful after handful of some yellow substance and heap it into the cavity he had made. Then he struck a match, lighted the stuff, and came sliding swiftly to earth again. We all drew off and waited.

"That," explained the bee-master, as he leaned on his woodman's axe out of breath, " is cotton-waste, soaked in creosote, and then smothered in powdered brimstone. See! it is burning famously. The fumes will soon fill the hollow of the tree and settle the whole company. Then we shall cut away enough of the rotten wood above to get all the best of the combs out; there are eighty pounds of good honey up there, or I'm no bee-man.   And then it's back to the clover-field for more guide-bees, and away on a new scent."

CHAPTER X

THE PHYSICIAN IN THE HIVE

IT was a strange procession coming up the red-tiled path of the bee garden. The bee-master led the way in his Sunday clothes, followed by a gorgeous footman, powdered and cockaded, who carried an armful of wraps and cushions. Behind him walked two more, supporting between them a kind of carrying-chair, in which sat a florid old gentleman in a Scotch plaid shawl; and behind these again strode a silk-hatted, black-frocked man carefully regulating the progress of the cavalcade. Through the rain of autumn leaves, on the brisk October morning, I could see, afar off, a carriage waiting by the lane-side; a big old-fashioned family vehicle, with cockaded servants, a pair of champing greys, and a glitter of gold and scarlet on the panel, where the sunbeams struck on an elaborate coat-of-arms.

The whole procession made for the extracting-house, and all work stopped at its approach. The great centrifugal machine ceased its humming. The doors of the packing-room were closed, shutting off the din of saw and hammer. Over the stone floor in front of the furnace—where a big caldron of metheglin was simmering—a carpet was hastily unrolled, and a comfortable couch brought out and set close to the cheery blaze.

And now the strangest part of the proceedings commenced. The old gentleman was brought in, partially disrobed, and transferred to the couch by the fireside. He seemed in great trepidation about something. He kept his gold eyeglasses turned on the bee-master, watching him with a sort of terrified wonder, as the old bee-man produced a mysterious box, with a lid of perforated zinc, and laid it on the table close by. From my corner the whole scene was strongly reminiscent of the ogre's kitchen in the fairy-tale; and the muffled sounds from the packing-room might have been the voice of the ogre himself, complaining at the lateness of his dinner.

Now, at a word from the black-coated man, the bee-master opened his box. A loud angry buzzing uprose, and about a dozen bees escaped into the air, and flew straight for the window-glass. The bee-master followed them, took one carefully by the wings, and brought it over to the old gentleman. His apprehensions visibly redoubled. The doctor seized him in an iron, professional grip.

" Just here, I think. Close under the shoulder-blade. Now, your lordship . . . "

Viciously the infuriated bee struck home. For eight or ten seconds she worked her wicked will on the patient. Then, turning round and round, she at last drew out her sting, and darted back to the window.

But the bee-master was ready with another of his living stilettos. Half a dozen times the operation was repeated on various parts of the suffering patient's body. Then the old gentleman—who, by this time, had passed from whimpering through the various stages of growing indignation to sheer undisguised profanity—was restored to his apparel. The procession was re-formed, and the bee-master conducted it to the waiting carriage, with the same ceremony as before.

As we stood looking after the retreating vehicle, the old bee-man entered into explanations.

" That," said he, " is Lord H——, and he has been a martyr to rheumatism these ten years back. I could have cured him long ago if he had only come to me before, as I have done many a poor soul in these parts; but he, and those like him, are the last to hear of the physician in the hive. He will begin to get better now, as you will see. He is to be brought here every fortnight; but in a month or two he will not need the chair. And before the winter is out he will walk again as well as the best of us."

We went slowly back through the bee-farm. The working-song of the bees seemed as loud as ever in the keen October sunshine. But the steady deep note of summer was gone; and the peculiar bee-voice of autumn—shrill, anxious, almost vindictive—rang out on every side.

" Of course," continued the bee-master, " there is nothing new in this treatment of rheumatism by bee-stings. It is literally as old as the hills. Every bee-keeper for the last two thousand years has known of it. But it is as much as a preventive as a cure that the acid in a bee's sting is valuable. The rarest thing in the world is to find a bee-keeper suffering from rheumatism. And if every one kept bees, and got stung occasionally, the doctors would soon have one ailment the less to trouble about."

" But," he went on, " there is something much pleasanter and more valuable to humanity, ill or well, to be got from the hives. And that is the honey itself. Honey is good for old and young. If mothers were wise they would never give their children any other sweet food. Pure ripe honey is sugar with the most difficult and most important part of digestion already accomplished by the bees. Moreover, it is a safe and very gentle laxative. And probably, before each comb-cell is sealed up, the bee injects a drop of acid from her sting. Anyway, honey has a distinct aseptic property. That is why it is so good for sore throats or chafed skins."

We had got back to the extracting house, where the great caldron of metheglin was still bubbling over the fire. The old bee-keeper relieved himself of his stiff Sunday coat, donned his white linen overalls, and fell to skimming the pot.

"There is another use," said he, after a ruminative pause, " to which honey might be put, if only doctors could be induced to seek curative power in ancient homely things, as they do with the latest new poisons from Germany. That is in the treatment of obesity. Fat people, who are ordered to give up sugar, ought to use honey instead. In my time I have persuaded many a one to try it, and the result has always been the same—a steady reduction in weight, and better health all round. Then, again, dyspeptic folk would find most of their troubles vanish if they substituted the already half-digested honey wherever ordinary sugar forms part of their diet. And did you ever try honey to sweeten tea or coffee? Of course, it must be pure, and without any strongly-marked flavour; but no one would ever return to sugar if once good honey had been tried in this way, or in any kind of cookery where sugar is used."

The bee-master ran his fingers through his hair, of which he had a magnificent iron-grey crop. The fingers were undeniably sticky; but it was an old habit of his, when in thoughtful mood, and the action seemed to remind him of something. His eyes twinkled merrily.

"Now," said he, "you are a writer for the papers, and you may therefore want to go into the hair-restoring business some day. Well, here is a recipe for you. It is nothing but honey and water, in equal parts, but it is highly recommended by all the ancient writers on beemanship. Have I tried it? Well, no; at least, not intentionally. But in extracting honey it gets into most places, the hair not excepted. At any rate, honey as a hair-restorer was one of the most famous nostrums of the Middle Ages, and may return to popular favour even now. However, here is something there can be no question about."

He went to a cupboard, and brought out a jar full of a viscid yellow substance.

"This," he said, "is an embrocation, and it is the finest thing I know for sprains and bruises. It is made of the wax from old combs, dissolved in turpentine, and if we got nothing else from the hives bee-keeping would yet be justified as a humanitarian calling. Its virtues may be in the wax, or they may be due to the turpentine, but probably they lie in another direction altogether. Bees collect a peculiar resinous matter from pine trees and elsewhere, with which they varnish the whole surface of their combs, and this may be the real curative element in the stuff."

Now, with a glance at the clock, the bee-master went to the open door and hailed his foreman in from his work about the garden. Between them they lifted away the heavy caldron from the fire, and tilted its steaming contents into a barrel close at hand. The whole building filled at once with a sweet penetrating odour, which might well have been the concentrated fragrance of every summer flower on the countryside.

" But of all the good things given us by the wise physician of the hive," quoth the old bee-keeper, enthusiastically, " there is nothing so good as well-brewed metheglin. This is just as I have made it for forty years, and as my father made it long before that. Between us we have been brewing mead for more than a century. It is almost a lost art now; but here in Sussex there are still a few antiquated folk who make it, and some, even, who remember the old methers—the ancient cups it used to be quaffed from. As an everyday drink for working-men, wholesome, nourishing, cheering, there is nothing like it in or out of the Empire.

CHAPTER XI

WINTER WORK ON THE BEE-FARM

THE light snow covered the path through the bee-farm, and whitened the roof of every hive. In the red winter twilight it looked more like a human city than ever, with its long double rows of miniature houses stretching away into the dusk on either hand, and its broad central thoroughfare, where the larger hives crowded shoulder to shoulder, casting their black shadows over the glimmering snow.

The bee-master led the way towards the extracting-house at the end of the garden, as full of his work, seemingly, as ever he had been in the press of summer days. There was noise enough going on in the long lighted building ahead of us, but I missed the droning song of the great extractor itself.

" No; we have done with honey work for this year," said the old bee-man. " It is all bottled and cased long ago, and most of it gone to London. But there's work enough still, as you'll see. The bees get their long rest in the winter; but, on a big honey-farm, the humans must work all the year round."

As we drew into the zone of light from the windows, many sounds that from afar had seemed incongruous enough on the silent, frost-bound evening began to explain themselves. The whole building was full of busy life. A furnace roared under a great caldron of smoking syrup, which the foreman was vigorously stirring. In the far corner an oil engine clanked and spluttered. A circular saw was screaming through a baulk of timber, slicing it up into thin planks as a man would turn over the leaves of a book. Planing machines and hammers and handsaws innumerable added their voices to the general chorus; and out of the shining steel jaws of an implement that looked half printing-press and half clothes-wringer there flowed sheet after sheet of some glistening golden material, the use of which I could only dimly guess at.

But I had time only for one swift glance at this mysterious monster. The bee-master gripped me by the arm and drew me towards the furnace.

" This is bee-candy," he explained, " winter food for the hives. We make a lot of it and send it all over the country. But it's ticklish work. When the syrup comes to the galloping-point it must boil for one minute, no more and no less. If we boil it too little it won't set, and if too much it goes hard, and the bees can't take it."

He took up his station now, watch in hand, close to the man who was stirring, while two or three others looked anxiously on.

" Time! " shouted the bee-master.

The great caldron swung off the stove on its suspending chain. Near the fire stood a water tank, and into this the big vessel of boiling syrup was suddenly doused right up to the brim, the stirrer labouring all the time at the seething grey mass more furiously than ever.

" The quicker we can cool it the better it is," explained the old bee-keeper, through the steam. He was peering into the caldron as he spoke, watching the syrup change from dark clear grey to a dirty white, like half-thawed snow. Now he gave a sudden signal. A strong rod was instantly passed through the handles of the caldron. The vessel was whisked out of its icy bath and borne rapidly away. Following hard upon its heels, we saw the bearers halt near some long, low trestle-tables, where hundreds of little wooden boxes were ranged side by side. Into these the thick, sludgy syrup was poured as rapidly as possible, until all were filled.

" Each box," said the bee-master, as we watched the candy gradually setting snow-white in its wooden frames, " each box holds about a pound. The box is put into the hive upside-down on the top of the comb-frames, just over the cluster of bees; and the bottom is glazed because then you can see when the candy is exhausted, and the time has come to put on another case. What is it made of? Well, every maker has his own private formula, and mine is a secret like the rest. But it is sugar, mostly—cane-sugar. Beet-sugar will not do; it is injurious to the bees.

" But candy-making," he went on, as we moved slowly through the populous building, " is by no means the only winter work on a bee-farm. There are the hives to make for next season; all those we shall need for ourselves, and hundreds more we sell in the spring, either empty or stocked with bees. Then here is the foundation mill."

He turned to the contrivance I had noticed on my entry. The thin amber sheets of material, like crinkled glass, were still flowing out between the rollers. He took a sheet of it as it fell, and held it up to the light. A fine hexagonal pattern covered it completely from edge to edge.

"This," he said, "we call super-foundation. It is pure refined wax, rolled into sheets as thin as paper, and milled on both sides with the shapes of the cells. All combs now are built by the bees on this artificial foundation; and there is enough wax here, thin as it is, to make the entire honeycomb. The bees add nothing to it, but simply knead it and draw it out into a comb two inches wide; and so all the time needed for wax-making by the bees is saved just when time is most precious—during the short season of the honey-flow."

He took down a sheet from another pile close at hand.

"All that thin foundation," he explained, "is for section-honey, and will be eaten. But this you could not eat. This is brood-foundation, made extra strong to bear the great heat of the lower hive. It is put into the brood-nest, and the cells reared on it are the cradles for the young bees. See how dense and brown it is, and how thick; it is six or seven times as heavy as the other. But it is all pure wax, though not so refined, and is made in the same way, serving the same useful, time-saving purpose."

We moved on towards the store-rooms, out of the clatter of the machinery.

" It was a great day," he said, reflectively, " a great day for bee-keeping when foundation was invented. The bee-man who lets his hives work on the old obsolete natural system nowadays makes a hopeless handicap of things. Yet the saving of time and bee-labour is not the only, and is hardly the most important, outcome of the use of foundation. It has done a great deal more than that, for it has solved the very weighty problem of how to keep the number of drones in a hive within reasonable limits."

He opened the door of a small side-room. From ceiling to floor the walls were covered with deep racks loaded with frames of empty comb, all ready for next season. Taking down a couple of the frames, he brought them out into the light.

" These will explain to you what I mean," said he. " This first one is a natural-built comb, made without the milled foundation. The centre and upper part, you see, is covered on both sides with the small cells of the worker-brood. But all the rest of the frame is filled with larger cells, and in these only drones are bred. Bees, if left to themselves, will always rear a great many more drones than are needed; and as the drones gather no stores but only consume them in large quantities, a super-abundance of the male-bees in a hive must mean a diminished honey-yield. But the use of foundation has changed all that. Now look at this other frame. By filling all brood-frames with worker-foundation, as has been done here, we compel the bees to make only small cells, in which the rearing of drones is almost impossible; and so we keep the whole brood-space in the hive available for the generation of the working bee alone."

"But," I asked him, "are not drones absolutely necessary in a hive? The population cannot increase without the male bees."

"Good drones are just as important in a bee-garden as high-mettled, prolific queens," he said; "and drone-breeding on a small scale must form part of the work on every modern bee-farm of any size. But my own practice is to confine the drones to two or three hives only. These are stationed in different parts of the farm. They are always selected stocks of the finest and most vigorous strain, and in them I encourage drone-breeding in every possible way. But the male bees in all honey-producing hives are limited to a few hundreds at most."

Coming out into the darkness from the brilliantly-lighted building, we had gone some way on our homeward road through the crowded bee-farm before we marked the change that had come over the sky. Heavy vaporous clouds were slowly driving up from the west and blotting the stars out one by one. All their frosty sparkle was gone, and the night air had no longer the keen tooth of winter in it. The bee-master held up his hand.

"Listen!" he said. "Don't you hear anything?"

I strained my ears to their utmost pitch. A dog barked forlornly in the distant village. Some night-bird went past overhead with a faint jangling cry. But the slumbering bee-city around us was as silent and still as death.

"When you have lived among bees for forty

years," said the bee-master, plodding on again, " you may get ears as long as mine. Just reckon it out. The wind has changed; that curlew knows the warm weather is coming; but the bees, huddled together in the midst of a double-walled hive, found it out long ago. Now, there are between three and four hundred hives here. At a very modest computation, there must be as many bees crowded together on these few acres of land as there are people in the whole of London and Brighton combined. And they are all awake, and talking, and telling each other that the cold spell is past. That is what I can hear now, and shall hear—down in the house yonder—all night long."

CHAPTER XII

THE QUEEN BEE: IN ROMANCE
AND REALITY

"QUEENS?" said the Bee-Master of Warrilow, as he filled his pipe with the blackest and strongest tobacco I had ever set eyes on; "queens? There are hundreds of hives here, as you can see; and there isn't a queen in any one of them."

He drew at the pipe until he had coaxed it into full blast, and the smoke went drifting idly away through the still April sunshine. We were in the very midst of the bee-garden, sitting side by side on the honey-barrow after a long morning's work among the hives; and the old bee-man had lapsed into his usual contemplative mood.

" 'Tis a pretty idea," he went on, "this of royalty, and a realm of dutiful subjects, and all the rest of it, in bee-life. But experience in apiculture, as with most things of this world, does away with a good many fine and fanciful notions. Now, the mother-bee in a hive, whatever else you might call her, is certainly not a queen, in the sense of ruling over the other bees in the colony. The truth is she has little or nothing to do with the direction of affairs. All the thinking and contriving is done by the worker-bees. They have the whole management of the hive, and simply look upon the queen as a much prized and carefully-guarded piece of egg-laying machinery, to be made the most of as long as her usefulness lasts, but to be thrown over and replaced by another the moment her powers begin to flag."

" No; there are no queens, properly so called, in bee-life," he continued. " All that belongs to the good old times when there were nothing but straw-skeps, and 'twas well-nigh impossible to get at the rights of anything; so the bee-keeper went on believing that honey was made out of starshine, and young bees were bred from the juice of white honeysuckle, which was all pretty enough in its way, even though it warn't true. But nowadays, when they make hives with comb-frames that can be lifted out and looked at in the broad light of day, folk are beginning to understand a power of things about bees that were dark mysteries only a while ago."

He puffed at his pipe for a little in silence. Far away over the great province of hives, the clock on the extracting-house pointed to half-past twelve; and, true to their usual time, the home-staying bees —the housekeepers and nurses and lately hatched young ones—were out for their midday exercise. The foragers were going to and fro as thickly as ever with their loads of pollen and water for the still cradled larvæ within; but now round every hive a little cloud of bees hovered, filling the sunshine with the drowsy music of their wings. The old bee-man took up his theme again presently at the point he had broken it off,

" If," said he, " you keep a fairly close watch on the progress of any one particular hive, from the time the first eggs appear in the combs early in January, 'tis very easy to see how the old false ideas got into general use. At first glance a bee-colony looks very much like a kingdom; and the single large bee, that all the others pay court to and attend so carefully, seems very like a queen. Then, when you look a little deeper and begin to understand more, appearances are still all in favour of the old view of things. The mother-bee seems, on the face of it, a miracle of intelligence and foresight. While, as far as you know, all other creatures in the world bring forth their young of both sexes haphazard, this one can lay male or female eggs apparently at will. You watch her going from comb to comb, and the eggs she drops in the small cells hatch out females, and those she puts in the larger ones are always males, or drones. More than that: she seems always to know the exact condition of the hive, and to be able to limit her egg-laying according to its need, or otherwise, of population; for either you see her filling only a few cells each day in a little patch of comb that can be covered with the palm of your hand, or she goes to work on a gigantic scale, and, in twenty-four hours, produces eggs that weigh more than twice as much as her whole body."

He got up now and began pacing to and fro, as was his custom when much in earnest over his bee-talk.

" Then," he went on, " to cap all, as the honey season draws on to its height, you are forced presently to realise that the queen has conceived and is carrying through a scheme for the good of her subjects that would do credit to the wisest ruler ever born in human purple. Every day of summer sunshine has brought thousands of young bees to life. The hive is getting overcrowded. Sooner or later one of two things must happen—either the increase of population must be checked, or a great party must be formed to leave the old home and go out to establish another one. Then it is that the mother-bee seems to prove beyond a doubt her wisdom and queenliness. She decides for the emigration; but as a leader must be found for the party, and none is at hand, she forms the resolve to head it herself. From that moment a change comes over the whole hive. Preparation for the coming event goes on fast and furiously, and excitement increases day by day. But the queen seems to forget nothing. A new ruler for the old realm must be provided to take her place when she is gone for ever; and now you see a party of bees set to work on something that fairly beggars curiosity. At first it looks exactly like an acorn-cup in wax hanging from the under-edge of the comb. Perhaps the next time you look the cup has grown to twice its original size; and now you see it is half full of a glistening white jelly. The next time, maybe, you open the hive, the acorn has been added to the cup; the queen-cell is sealed over and finished, and about a week later there comes out a full-grown queen bee, twice the size of the ordinary worker and quite different in shape and often in colour too. But days before the new ruler is ready the excitement in the hive has grown to fever-pitch. If you come out then in the quiet of the night and put your ear close to the hive, you will hear a shrill piping noise which the ancient skeppists tell you is the old queen calling her subjects together for the swarm on the morrow. And, sure enough, out she goes with half the population of the hive in her train, to look for a new home; and in a day or so the new queen comes out of her cell to take charge of the colony."

He paused to fill the old briar pipe again, lighting it with slow deliberate puffs, and I could not help marking how nearly alike in colour were the bowl and his rugged, sunburnt, clever face.

"But now, look you!" said he, suddenly levelling the pipe-stem like a pistol at me to emphasise his words. "If the mother-bee really brought all this about, queen would not be a good enough name for her. But the truth is, throughout all the wonder-workings of the hive, the queen is little more than an instrument, a kind of automaton, merely doing what the workers compel her to do. They are the real queens in the hive, and the mother-bee is the one and only subject. Did you ever think what a queen-bee actually is, and how she comes to be there at all? The fact is that the workers have made her for their own wise purposes, just as they make the comb and the honey to store in it. The egg she is hatched from is in no way different from any worker-egg. If you take one from a queen-cell and put it in the ordinary comb, it will hatch out a common female worker-bee: and an egg transferred from worker-comb to a queen-cell becomes a full-grown queen. Thousands and thousands of worker-eggs are laid in a hive during the season, and each of those could be made into a queen if the workers chose. But the worker-egg is laid into a small cell, and the larva is bred on a bare minimum of food, at the least possible cost in time, trouble, and space to the hive; while, when a new queen is wanted, a cell as big as your finger-top is built, and the larva is stuffed like a prize-pig through all its five days of active life, until, with unlimited food and time and room to grow in, it comes out at last a perfect mother-bee.''

" But,'' I asked him, " how is the population in the hive regulated, and how can the apportionment of the sexes be brought about? If, as you say, the queen does only what she is made to do by the workers, and that unthinkingly and mechanically, you only increase the difficulty of the problem.''

" As for increasing or restricting the number of eggs laid,'' he said, " that is only a question of food; and here you see how the workers control the mother-bee entirely, and, through her, the whole condition of the hive. When she is egg-laying they feed her from their own mouths with special predigested food; and the more she gets of this, the more eggs are laid. But when the season is done, and the need for a large population over, this rich stimulating diet is kept from her. She then must go to the honey-cells like the rest, or starve; and at once her egg-laying powers begin to fall off. And it is in exactly the same way—by their management of the queen—that the workers control the proportion of the sexes in a hive. 'Tis more difficult to explain, but here is about the rights of it. Directly the new-hatched queen-bee is ready for work, she flies out to meet the drones; and one impregnation lasts her whole life through. But the eggs themselves are not fertilised until the very moment of laying, and then only in the case of those laid in worker-comb: drone-eggs are never impregnated at all. Now, in all likelihood, as the queen is being driven over the combs, it is the size of the cell that determines whether the egg laid shall be male or female. When the queen thrusts her long pointed body into the narrow worker-cell, her position is a straight, upright one, and the egg cannot be laid without passing over the impregnation-gland; but with the larger drone-cell the queen has room to curve herself, which is the means, I think, of the egg escaping without being fertilised. And so you see it is only the female bee that has two parents; the drone has no father at all."

CHAPTER XIII

THE SONG OF THE HIVES

FROM the lane, where it dipped down between its rose-mantled hedges, nothing of the bee-garden could be seen. The dense barricade of briar and hawthorn hid all but the lichened roof of the ancient dwelling-house; and strangers going by on their way to the village saw nothing of the crowding hives, and marked little else than the usual busy murmur of insect-life common to any sunny day in June.

But when they came out of the green tunnel of hedgerows into the open fields beyond, chance wayfarers always stopped and looked about them wonderingly, at length fixing a puzzled glance intently on the blue sky itself. At this corner, and nowhere else, seemingly, the air was full of a deep, reverberant music. A steady torrent of rich sound streamed by overhead; and yet, to the untutored observer, the most diligent scrutiny failed to reveal its origin. A few gnats harped in the sunbeams. Now and again a bumble-bee struck a deep chord or two in the wayside herbage underfoot. But this clear, strong voice from the skies was altogether unexplainable. To human sight, at least, the blue air and sunshine held nothing to account for it; and the stranger unversed in honey-bee lore, after taking his fill of this melodious mystery, generally ended by giving up the problem as insoluble, and passing on to his business or pleasure in the little green-garlanded hamlet under the hill.

That the bees of a fairly large apiary should produce a considerable volume of sound in their passage to and fro between the hives and the honey-pastures is in no way remarkable. In the heyday of the year—the brief six weeks' honey-flow of the English summer—probably each normal colony of bees would send out an army of foragers at least twenty thousand strong. What really seems matter for wonder is the way in which bees appear to concentrate their movements to certain well-defined tracks in the atmosphere. They do not distribute themselves broadcast over the intervening space, as they might be expected to do, but wonderfully keep to certain definite restricted thoroughfares, no matter how near or how remote their foraging grounds may be.

And this particular gap in the chain of hedgerows really marked the great main highway for the bees between the hives and the clover-fields silvering the whole wide stretch of hill and dale beyond. Every moment had its winged thousands going and returning. At any time, if a fine net could have been cast suddenly a few fathoms upward, it would have fallen to earth black and heavy with bees; but the singing multitude went by at so fast and furious a pace that, to the keenest sight, not one of the eager crew was visible. Only the sound of their going was plain to all; a mighty tenor note abroad in the sunshine, a thronging sustained melody that never ceased all through the heat and burthen of the glittering summer's day.

When Shelley heard the " yellow bees in the ivy-bloom," and he of Avonside wrote of " singing masons building roofs of gold," probably neither thought of the humming of the hive-bee as anything more than an ingredient in the general delightful country chorus, as distinct from the less-inspiring labour-note of busy humanity in a town. With the single exception, perhaps, of Wordsworth, poets, thinking most of their line, commonly miss the subtler phases of wild life, such as the continually changing emphasis and capricious variation in bird song, the real sound made by growth, or the unceasing movement of things conventionally held to be inert. And in the same way the endlessly varied song of the bees has been epitomised by imaginative writers generally into a sound, pleasantly arcadian enough, but little more suggestive of life and meaning than the hum of telegraph wires in a breeze.

Yet there are few sounds in nature more bewilderingly complex than this. For every season in the year the song of the hives has its own distinct appropriate quality, and this, again, is constantly influenced by the time of day, and even by the momentary aspect of the weather. A bee-keeper of the old school—and he is sure to be the " character," the quaint original of a village—manages his hives as much by ear as by sight. The general note of each hive reveals to him intuitively its progress and condition. He seems to know what to expect on almost any day in the year, so that if Rip van Winkle had been an apiarist the nearest bee-garden would have been as sure a guide to him, in respect of the time of year at least, as the sun's declining arc in the heaven is to the tired reapers in respect of the hour of day.

Most people—and with these must be included even lifelong country-dwellers—are wont to regard the humming of the hive-bee as a simple monotone, produced entirely by the rapid movement of the wings. But this conception halts very far short of the actual truth. In reality, the sound made by a honey-bee is threefold. It can consist either of a single tone, a combination of two notes, or even a grand triple chord, heard principally in moments of excitement, such as when a swarming-party is on the wing, or in late autumn and early spring, when civil war will often break out in an ill-managed apiary. The actual buzzing sound is produced by the wings; the deeper musical tones by the air alternately sucked in and driven out through the spiracles, which are breathing-tubes ranged along each side of a bee's body; while the shrill, clarinet-like note comes from the true voice-apparatus itself. In ordinary flight it is the wings and the respiration-tubes conjointly which produce the steady volume of sound heard as the honey-makers stream over the hedgetop towards the distant clover-fields; and this is the note also that pervades the bee-garden through every sunny hour of the working-day. The rich, soft murmur coming from the spiracles is probably never heard except when the bee is flying, but both the true voice and the whirring wing-melody are familiar as separate sounds to every bee-keeper who studies his hives.

When the summer night has shut down warm and still over the red dusk of evening, and the last airy loiterer is safely home from the fields, a curious change comes to the bee-garden. The old analogy between a concourse of hives and a human city is, at this season, utterly at fault. Silence and rest after the day's work may be the portion of the larger community, but in the time of the great honey-flow there is neither rest nor slumber for the bees. A fury of labour possesses them, one and all; and darkness does not remit, but merely transposes the scene of, their activity. Coming out into the garden at this hour for a quiet pipe among the hives—an old and favourite habit with most bee-keeping veterans—the new spirit abroad is at once manifest. The sulky, fragrant darkness is silent. quiet with the influence of the starshine overhead; but the very earth of the footway seems to vibrate with the imprisoned energy of the hives. This is the time when the low, rustling roar of wing-music can best be heard, and one of the most wonderful phases of bee-life studied. The problem of the ventilation of human hives is attacked commonly on one main principle—unstinted ingress for fresh air and a like abundant means of outward passage for the bad. But, if the bees are to be credited, modern sanitary scientists are trimming altogether on the wrong tack. A colony of bees will allow one aperture, and one alone, in the hive, to serve all and every purpose. If the enterprising novice in beemanship gimlets a row of ventilation-holes in the back of his hive—an idea that occurs to most tyros in apiculture—the bees will infallibly seal them all up again before morning. They work on entirely different principles, impelled by their especial needs. The economy of the hive requires the temperature to be absolutely and immediately within the control of the bees, and this is only possible when the ventilatory system is entirely mechanical. The evaporation of moisture from the new-gathered nectar, and the hatching of the young brood, necessitate an amount of heat much less than that required for wax-generating; as soon as the wax-makers begin to cluster the temperature of the hive is at once increased. But if a current of air were continually passing through the hive these necessary heat variations would be difficult to manage, even supposing them possible at all; so the bees have invented their unique system of a single passage-way, combined with an ingenious and complicated process of fanning, by which the fresh air is sucked in at one side of the entrance and the foul air drawn out at the other, the atmosphere of the hive being thus maintained in a constant state of circulation, fast or slow, according to the temperature needed.

In the hot summer weather these fanning-parties are at work continuously, being relieved by others at intervals of a few minutes throughout the day. But at night, when the whole population of the hive is at home, the need for ventilation is greatly augmented, and then the open lines of fanners often stretch out over the alighting-board six or seven ranks deep, making an harmonious uproar that, on a still night, will travel incredible distances.

This tense, forceful labour-song of the bee-garden, heard unremittingly throughout the hours of darkness, is always pleasant, often indescribably soothing in its effect. But it is essentially a communal note, expressive only of the well or ill being of the hive at large. The individuality, even personal idiosyncrasy, which undoubtedly exists among bees, finds its utterance mainly through the true voice-organ. You cannot stand for long, here, in the quiet of the summer night, listening to one particular hive, without sooner or later becoming aware of other sounds, in addition to the general musical hubbub of the fanning army. It is evident that a nervous, high-strung spirit pervades the colony, especially during the season of the great honey-flow. Their common agreement on all main issues does not prevent these " virgin daughters of toil " from engaging in sundry sharp altercations and mutual hustlings in the course of their business; and, at times of threatening weather, a tendency towards snappishness, and a whimsical perversity characteristically feminine, seem to make up the prevailing tone. It is during these chance forays that the true voice of the honey-bee, apart from the sounds made by wing and spiracle, can best be differentiated.

CHAPTER XIV

CONCERNING HONEY

THE bee-keepers in English villages to-day are all familiar—too familiar at times—with the holiday-making stranger at the garden gate inquiring for honey. Somehow or other the demand for this old natural sweet-food appears to have greatly increased of recent years among wandering towns-folk in the country. A competent bee-master, dealing with a large number of combs, will not mingle them indiscriminately, but will unerringly assort them, so that he will have perhaps at the end of the season almost as many kinds of honey in store as there are fields on his countryside. I speak, of course, not of the large bee-farmer—who, em-ploying of necessity wholesale methods, can aim only at a good all-round commercial sample of no finely distinctive colour or flavour—but of the con-noisseur in bee-craft, the gourmet among the hives, who knows that there are as many varieties in honey as there are in wine, and would as little dream of confusing them.

Honey lovers who have been eating wax all their days will be as hardly dissuaded from the practice as he whose custom it may be to consume the paper in which his butter is wrapped, or take a proportion of the blue sugar-bag with the lumps in his tea. Yet the last are no more absurdities than the former, except in degree. Pure beeswax has neither savour nor nutrient properties, and passes wholly unassimilated through the human system. Even the bees themselves cannot feed upon it when at dire extremes: the whole hive may die of starvation in the midst of waxen plenty. Of all creatures, mice, and the larvæ of two species of moth, alone will make away with it; and even in their case it is doubtful whether the comb be not destroyed for the sake of the odd grains of pollen and the pupa-skins it contains. Broadly speaking, unless you can trust a dipped finger-tip to reveal to you on the moment the qualities of this village-garden honey, it is always safer to buy in the comb. But the wax should never be eaten. The proper way to deal with honeycomb at table is to cut it to the width of the knife-blade; and, laying it upon the plate with the cells vertical, press the blade flat upon it, when the honey will flow out right and left. In this way, if duly carried out, the honey is scientifically separated, no more than one per cent remaining in the slab of wax.

The Bee as a Chemist

It is not strange, because it is so common, to find people who have eaten honeycomb regularly all their lives, yet are unknowingly ignorant of the first rudimentary fact in its nature and composition. To know that you do not know is an intelligible state, the initial true step towards knowledge; but to be full of erroneous information, and that complacently, is to be ignorant indeed. Of such are the old lady who dwelt in the Mile End Road, and believed that cocoanuts were monkeys' eggs, and the man who will tell you without expectancy of contradiction that honey is the food of bees.

Now this is no essay in cheap paradox, but a sober attempt to reinstate in the public mind the unsophisticated truth. The natural foods of the bee-hive are the nectar and the pollen, the " love ferment " of the flowers. On these the bee subsists entirely, so long as she can obtain them, and will go to her honey stores only when nature's fresh supplies have failed. One speaks by poetic licence, or looseness, of bees gathering honey from blossoming plants. The fact is they do nothing of the kind, and never did. The sweet juices of clover, heather, and the like, differ fundamentally, both in appearance and in chemical properties from honey. Though the main ingredient in honey is nectar, the two are totally different things; and honey, far from being the normal food of bees, is only a standby for hard times, a sort of emergency ration, put up in as little compass and with as great a concentration as such things can be.

The story of how honey is made, and why it is made at all, forms one of the most interesting items in the history of the hive-bee. In a land where nectar-yielding plants flourish all the year through, if such a spot exist at all, there would be no honey, because the necessity for it would not occur. Hive-bees in such a land would go all their lives, and assuredly never dream of honey-making. But wherever there is winter, or a season when the supply of nectar and pollen temporarily fails, the bee, who does not hibernate in the common sense of the term, must devise a means of supporting life through the famine period. Many creatures can and do accomplish this by merely laying up in a comatose condition until such time as their natural food is plentiful again, and they may safely resume their old activities. But this will not do for the doughty honey-bee. A curious aspect of her life is the way in which she appears to recognise the competitive spirit in all the higher forms of earthly existence, and deliberately sets herself in the fore-rank of affairs with that principle in view. It would be easy for a few hundred worker-bees to get together in some warm nook underground, with that carefully tended piece of egg-laying mechanism, their queen, in their midst; and in a semi-dormant condition to pass the dark winter months through, gradually rousing their own fires of life as the year warmed up again in the spring. But such a system would mean that the colony would have to start afresh from the bottom of the ladder of progress with every year. The hive-bee has conceived a better plan, and the basis, the essential factor of it all, is this thing of mystery which we call honey.

The True Purpose of the Hive

The ancient Roman name for a beehive was *alvus*, which, translated into its blunt Anglo-Saxon equivalent, means belly. And this gives us in a word the whole secret about honey-making. As a matter of fact, the hive in summer acts as a digestive chamber, wherein the winter aliment of the stock is prepared. The bees, during their ordinary work-aday life, subsist on the nectar and pollen which they are continually bringing into the hive. Much pollen is laid by in the cells in its raw condition, but pollen is almost exclusively a tissue-former, and it is not used by the worker-bees during the winter for their own sustenance, but preserved until early spring, when it forms the principal component in the bee-milk on which the larvæ are mainly fed. The nectar, however, is necessary at all times to support life in the mature bees, and it must therefore be stored for use during the long months when there are no flowers to secrete it.

It is here that we get a glimpse into the ways of the honey-bee that may well give spur to the most wonder-satiated amongst us. If a sample of fresh nectar is examined, it will be found to consist of about seventy per cent of water, the small remainder of its bulk being made up of what is chemically known as cane sugar, together with a trace of cer-tain essential oils and aromatic principles. It is practically nothing but sweetened and flavoured water. But ripe honey shows a very different composition. The oils and essences are there, with some added acids; but of water there is no more than seven to ten per cent; practically the entire bulk of good honey consists of sugar, but it is grape sugar, with scarce a trace of the cane sugar which nectar exclusively contains. To put the thing in plainest words—the economic honey-bee, finding herself with three or four months to get through at the least possible cost in energy and nutriment, has scientifically reasoned out the matter, and, among other ingenious provisions, has arranged to subject her winter food to a process of pre-digestion during the summer, so that when she consumes it there shall be neither force expended in its assimilation nor waste products taken with it, needing to be afterwards expelled. Honey, in fact, is the nectar digested, and then regurgitated just when it is ready to be absorbed into the system. It is almost certain that every drop goes through this process twice, and possibly three times, in each case by different bees; and the heat of the hive still further contributes to the object in view by driving off the superfluous moisture from the nectar so treated, and thus concentrating it into an almost perfect food.

CHAPTER XV

IN THE ABBOT'S BEE-GARDEN

STANDING in the lane without, and looking up at the grey forbidding walls of the old abbey, you wondered how anything human could exist on the other side; but, once past the heavy iron-studded gate, your thoughts doubled like hares in the opposite direction.

It seemed good to be a monk, if life could be all sunshine, and quietude, and beauty like that. As you waited in the shadow of the great stone-flagged portico, while your coming was announced, this feeling grew deeper with every moment. The garden sloped down to the river's edge, winding footway, and green lawn, and kitchen-plot all alike girdled and barricaded with rich-hued autumn flowers. Through the mass of crimson fuchsia and many-coloured dahlia and hollyhock, bowers of pink and white geranium with stems as thick as your wrist, ancient apple-trees drooping under their burden of scarlet fruit, crowding jungles of roses, you could see the bright waters sweeping by, and hear their busy sound as they won a way amidst the rocky boulders strewing the bed of the tortuous Devon stream.

Here and there in the sunny field-of-view visible through the arched doorway, black-robed figures were quietly at work: some digging; others gathering apples in the orchard; one sturdy brother was mowing the Abbot's lawn, the bright blade coming perilously near his fluttering skirts at every stroke; another went by trundling a wheelbarrow full of green vegetables for the refectory table. There was a distant cackle of poultry, blending oddly with the solemn chant that came from the chapel hard by. Robins sang everywhere, and starlings clucked and whistled in the valerian that topped the great encircling wall. But wherever you looked, whatever drew away your attention for the moment, you were sure to come back to the consideration of one preponderant yet inexplicable thing. A steady, deep note was upon the air. Rich and resonant, it seemed to come from all directions at once. The dim, grey-vaulted entrance-porch was full of it. Looking up into the dusk of oaken beams overhead, there it seemed at its strangest and loudest. Queerest fact of all, it appeared to have some mysterious affinity with the sunshine, for when a stray white argosy of cloud came drifting over the azure and obscured for a minute the glad light, this full, sonorous note died suddenly away, rising as swiftly again to its old power and volume when the sunbeams glowed back once more over the spacious garden, and over the riverside willows that shed their gold of dying leafage with every breath of the soft south wind.

It was not until you stepped outside, and looked

upward over the face of the old building, that you realised what it all meant. From its foundation to the highest stone of the ancient bell-turret, the whole front of the place was thickly mantled with ivy in full flower, and every yellow tuft of blossom was besieged with bees. There seemed tens of thousands of them, hovering and humming everywhere; and thousands more arriving with every moment out of the blue air, or darting off again fully laden, and away to some invisible bourne over the ruddy roof of orchard trees.

Intent on this vociferous wonder, you do not catch the footfall on the gravel-path in your rear, or see the sombre figure of the Abbot as he comes towards you, the sweep of his black frock setting all the marigolds nodding behind him, as though from a sudden flaw of wind. And now you have another pleasurable disillusionment as to monkish conditions of being. Trudging along the deep-cut Devonshire lanes on your way to the Abbey, through the rain of falling autumn leaves, you pictured the place to yourself as a kind of sacred sink of desolation, inhabited by a crew of sour-visaged anchorites, who found only godlessness in sunshine, and in cakes-and-ale nothing but assured perdition. But here, coming towards you, smiling, and with outstretched hand, is the last kind of human being you expected to see. Clad from head to foot in sober black, with, for ornament, but the one plain silver cross swinging at his breast, the Abbot shows, unmistakably, for a gentleman of cultured and enlightened mien. A fine, swarthy face, kind, calm eyes behind gold spectacles, a voice like an old violin, and a grip of the hand that makes you wince with its abounding welcome, all combine to set you there and then at your ease; and talk begins at once on the old, familiar plane among bee-keepers—the quick, enthusiastic interchange, each participant as ready a listener as learner, common all the world over, wherever flowers grow and men love bees.

The brothers of the old Benedictine monastery —so the Abbot tells you, as he leads the way towards the hives, through the sun-riddled labyrinth—have kept bees, probably, for more than a thousand years. There is no doubt that the original abbey building stood there, in the wooded cleft of Devon valley, so long ago as the sixth century, nor little question that its founder was a bee-man, for he was contemporary and friend of the great St Modonnoc who himself first taught Irishmen to keep bees.

"Monks, in the very earliest times, were almost invariably apiculturists," argues the Abbot. He stops in the orchard, the more impressively to quote Latin, the glib leaf-shadows playing the while over his tonsured head. "*Lac et mel*; *panis, vena rudis*. Milk and honey, and coarse oaten bread. At least we know, from our chronicles, that these were the common daily fare of our Order more than eight hundred years ago; and honey remains a part of our food to this day."

Thus overawed with the centuries, you begin to form a mental picture of the bee-garden you are about to visit, voyaging so pleasantly through winding path and shady thicket, with the bell-like sound of the water growing clearer and clearer at every step. With all that hoary tradition of the ages

behind them, you promise yourself, these monks
will have clung to their bee-keeping mediævalism
as to some sacred, inviolable thing. There will be
no movable comb-frames, nor American sections,
nor weird, foreign races of bees. They will never
have heard even of foul-brood, or napthol-beta, or
the host of things that bless or curse modern apiculture at every turn of the way. But, instead, there
will be a tangled wilderness of late blossom, such
as only Devonshire can show in November; dome-
shaped hives of straw, each with its singing company about it; perhaps a superannuated brother or
two quietly making straw hackles to shield the hives
against coming winter weather; even, perchance,
the smell of burning brimstone on the air, as the
last remnant of the honey-harvest is gathered in the
ancient way, by " taking up " the strongest and
the weakest colonies of bees.

And then a wicket-gate in the old wall determines
the path and your ruminations together. A sudden
burst of sunshine; the rich medley of sound from
fourscore hives lifting high above the song of the
purling stream; and you are out on the broad, green
river-bank, looking on at a scene very different from
the one you have expected.

There are no old-fashioned hives; they are all of
the latest, most scientific pattern, ranged under the
shelter of the wall in two wide terraces of close-
shaven turf, looking southward over the stream.
There are outhouses of the most approved design,
where all the business of a modern apiary is going
on. Here and there you see black-frocked figures
at work, dexterously examining the colonies. There
is the deep, whirring note of honey-extractors; the

clamour of carpenters' tools; the faint, sickly smell from the wax-boilers; all the familiar evidences of bee-farming carried on in the most modern, twentieth-century way.

As you look down the long, trim avenue of gaily-painted hives your companion has a quiet side-glance upon you, obviously noting your disappointment.

"What would you?" says he, and his deep voice rings like a passing-bell for all your dreams. "Everything must move with the times, or must inevitably perish. Modernism, rightly understood, is God's fairest, most priceless gift to the universe. It is a crucible through which all things of true metal must pass to lose the accumulated dross of the ages, keeping their original pure substance, but taking the new shape required of them by latter-day needs. It is so with the old, dim windows of man's faith; daily the glass is being taken out, smelted down, purified, replaced; we can see abroad into distances now never before visible. And so it must prove even with bee-keeping, which is one of the oldest human occupations in the world."

He waves his hand towards the sunny prospect before you. Beyond the river the burning apple-woods soar steadily upward; and high above these, stretching away to meet the blue sky, lie the Devon moorlands, once all rose-red with blossoming heather, but now, parched and brown, except where a grey crag or rock puts forth its jagged head.

"It is a fine thing, perhaps," says the Abbot, thoughtfully swinging his silver cross in the sun-beams, "to love old, ignorant customs, old, benighted, useless errors, for their picturesqueness and beauty alone. But don't you think it is a still finer thing to teach poor people how they may win from the common hillside plenty of rich, nourishing food at almost no cost at all? And that is what we are doing here. Modern bee-science, it is true, gives us only an ugly utilitarian hive. It sweeps away all the bright, iridescent cobwebs in the path of bee-keeping, and substitutes hard fact for pretty fairy-tale. But the sum of it all is that the poor cottager gains, not twenty or thirty pounds at most of coarse, unsaleable sweet food from his hives, but perhaps hundredweights of pure, choice, section-honey, which, sold in the proper market, will clothe his children comfortably, and make it possible for them to lead decent human lives."

CHAPTER XVI

BEES AND THEIR MASTERS

THERE are three great tokens of the coming of spring in the country—the elm-blossom, the cry of the young lambs, and the first rich song of the awakening bees.

All three come together about the end of February or beginning of March, and break into the winter dearth and silence in much the same sudden, unpremeditated way. You look at the woodlands, cowering under the lash of the shrill north wind, and all seems bare and black and lifeless. But the wind dies down in a fiery sunset. With the darkness comes a warm breath out of the west. On the morrow the spring sunshine runs high through all the valleys like liquid gold; the elm-tops are ablaze with purple; from the lambing-pens far and near a new cry lifts into the still, warm air; and in the bee-gardens there is the unwonted, old-remembered symphony, prophetic of the coming summer days.

The shepherd, the bee-man, the woodlander—these three live in the focus of the seasons, and feel their changes long before any other class of country folk. But the bee-man, if he would prosper, must take the sun as his veritable daily guide from year's end to year's end. Those whose conception of a bee-keeper is mainly of one who looks on from his cottage door while his winged thousands work for him, and who has but to stretch out his hand once a year to gather the hoard he has had no part in winning, know little of modern beemanship. This would be almost literally true of the old skeppist days, when bees were left much to their own devices, and thirty pounds of indifferent honey was reckoned a good take from a populous hive. But the modern movable comb-frame has altered all that. Now ninety or a hundred pounds weight of honey per hive is expected, with ordinarily good seasons, on a well-managed bee-farm; and in exceptional honey-flows very strong stocks of bees have been known to double and even treble that amount.

The movable comb-frame has three prime uses. The hives can be opened at any time and their condition ascertained without having to wait for outside indications. Brood-combs, with the young bees all ready to hatch out, can be taken from strong colonies and given to weak ones, and thus the population of all stocks may be equalised. The filled honeycombs can be removed, emptied by the centrifugal extractor, and the combs returned to the hive ready for another charge; and so the most onerous and exacting labour of the hive, comb-building, is largely obviated.

The modern beehive has another great advantage over the old straw skep, in that its size can be regulated according to the needs of each colony. More combs can be added as the stock grows, and thus no limit is set to its capacity. With the ancient form of hive fifteen or twenty thousand bees meant a crowded citadel, and there was nothing for it but to relieve the congestion by swarming. But the swarming habit has always been the principal obstacle to large honey-takes; and the problem which the modern bee-keeper has to solve is how to prevent his stocks from thus breaking themselves up into several hopelessly weak detachments.

It is all a war of wits between the bees and their masters. In nature the honey-bee is possessed of an inveterate caution. Famine is especially dreaded, and the number of mouths to fill in a hive is always kept strictly to the limits of the incoming food-supply. Thus a natural bee-colony is seldom ready for the honey-flow when it begins in early April, because it is only then that the raising of the young brood is allowed its fullest scope. This, however, is of no importance as far as the bees themselves are concerned, for a balance of stores of about twenty pounds weight at the end of a season will safely carry the most populous colony through any ordinary winter.

But from the bee-master's point of view it means practically a lost harvest. All the arts and devices of the modern bee-keeper, therefore, are set to work to overcome this timid conservatism of the hives, and to induce the creation of immense colonies of worker-bees as early as possible in the season, so that there may be no lack of labourers when the harvest is ready.

These first warm days of March, that bring the elm-blossom, and the cry of the lambs, and the old sweet music of the bee-gardens together, really

form the most critical time of all for the apiarist
who depends on his honey for his bread-and-butter.
It is the natural beginning of the bee-year, and on
his skill as a craftsman from now onward all chance
of a prosperous season will rest. It is true that,
within the hive, the bees have been awake and stirring for a long time past. Ever since the " turn of
the days," just before Christmas, the queen-mother
has been busy; and now there are young bees, little
grey fluffy creatures, everywhere in the throng; and
the area of sealed brood-cells is steadily growing.
But it is only now that the world out-of-doors becomes of any interest to the bees.

This is the time when the scientific bee-man must
get to work. His whole policy is one of benevolent
fraud. He knows that the population in his hives
will not be allowed to increase until there is a
steady, assured income of nectar and pollen. He
cannot create an early flower-crop, but he does
almost the same thing. Every hive is supplied with
a feeding-stage, where cane-sugar syrup, of nearly
the same consistency as the natural flower-secretion,
is administered constantly; and he places trays full of
pea-flour at different stations amongst his hives, as
a substitute for pollen. There is a special art in the
administration of this sugar-syrup. One might
think that if the bees required feeding at all, the
more they were given the better they would thrive.
But experience is all against this notion. The
artificial food is given, not to replenish an exhausted
larder, but to simulate a natural new supply. This,
in the ordinary state of things, would begin in about
a month's time, coming at first scantily, and gradually
increasing. By syrup-feeding early in March, the

bee-master sets the clock of the year forward by many weeks. He imitates nature by arranging his feeding-stages so that the supply of syrup can be limited to the actual day-to-day wants of the colony, allowing the bees freer access to the syrup-bottles from time to time as their numbers augment.

If this is adroitly done, the effect on the colony is remarkable. The little company of bees whose part it is to direct the actions of the queen-mother, seeing what is apparently the natural fresh supply of food coming in, in daily increasing quantities, at length cast their hereditary reserve aside, and allow the queen fullest scope for egg-laying. The result is that by the time the real honey-flow commences the population of each hive is double what it would be if it had been left to its own resources, and the honey-yield is more than proportionately great. It is well know among bee-men that a hive containing, say, forty thousand workers will produce very much more honey than two hives together numbering twenty thousand each.

There is another vital consideration in this work of early stimulation of the hives, which the capable bee-master will never neglect. When the natural honey-glut is on, the whole hive reeks with the odours given off from the evaporating nectar. The raw material, as gathered from the flowers, must be reduced by the heat of the hive and other agencies to about one-quarter of its original bulk before it is changed into mature honey. The artificial food given to the bees will, of course, have none of this scent, and the old honey-stores in the hive are hermetically sealed under their waxen cappings. To complete the deception which has been so elaborately contrived, the bee-master must furnish his hives with a new atmosphere. This he does by slicing off the cappings from some of the old store-combs, thus letting out their imprisoned fragrance, and filling the hive at once with the very essence of the clover-fields where the bees worked in the bygone summer days. The smell of the honey at this time, combined with the regular and increasing supply of syrup, acts like a powerful stimulant on the whole stock, and the work of brood-raising goes rapidly forward.

In intensive culture of all kinds there are risks to be run peculiar to the artificial state of things engendered, and modern bee-breeding is no exception to the rule. When once this fictile prosperity is installed by the bee-master, no lapse or variation in the due amount of food must occur. Even a single day's remission of supplies may undo all that a month's careful manipulation has brought about. English bees understand their native climate only too well, and the bitter experience of former years has taught them to be prepared for a return of hard weather at any moment. Under natural conditions, if a few weeks' warmth has induced them to raise population, and a sudden return of cold ensues, the bees will take very prompt and stern measures to meet the threatening calamity of starvation. The queen will cease laying at once; all unhatched brood will be ruthlessly torn from its cradle-cells and destroyed; old, useless bees will be expelled from the colony. And this is exactly what will happen if the artificial food-supply is allowed to fail even for the shortest period.

CHAPTER XVII

THE HONEY THIEVES

WHERE the bee-garden lay, under its sheltering crest of pine-wood, the April sunbeams seemed to gather, as water gathers in the lap of enclosing hills. Out in the lane the sweet hot wind sang in the hedgerows, and the white dust lifted under every footfall and went bowling merrily away on the breeze. But once among the crowding hives, you were launched on a still calm lake of sunshine, where the daffodils hardly swayed on their slender stems; and the smoke from the bee-master's pipe, as he came down the red-tiled path, hung in the air behind him like blue gossamer spread to catch the flying bees.

As usual, the old bee-man had an unexpected answer ready to the most obvious question.

" When will the new honey begin to come in? " he said, repeating my inquiry. " Well, the truth is honey never comes into the hives at all; it only goes out. That's the old mistake people are always falling into. Good bees never gather honey: they leave that to the wicked ones. If I had a hive of bees that took to honey-gathering, I should have to stop them, or end them altogether. It would have to be either kill or cure."

He took a quiet whiff or two, enjoying the effect of this seeming paradox, then went on to explain.

" What the bees gather from the flowers," said he, " is no more honey than barley and hops are beer. Honey has to be manufactured, first in the body of the bee, and then in the comb-cells. It must stand to brew in the heat of the hive, just as the wort stands in the gyle-tun; and when it is ready to be bunged down, before the bee adds the last little plate of wax to the cell-capping, she turns herself about and, as I believe, injects a drop of the poison from her sting—or seems to do so. Then it is real honey, but not before. Now, about these bad bees, the honey-gatherers——"

He stopped, putting his hand suddenly to his face. A bee had unexpectedly fastened her sting into his cheek. At the same moment another came at me like a spent shot from a gun, and struck home on my own face. The old bee-man took a hurried survey of his hives.

" Why," said he, " as luck, or ill-luck, will have it, I think I can show you the honey-gatherers at work now. There's only one thing that would make my bees wild on such a morning as this; and we must find out where the trouble is, and stop it."

He was looking about him in every direction as he spoke; and at last, on the farther side of the bee-garden, seemed to make out something amiss. As we passed between the long rows of bee-dwellings every hive was the centre of its own thronging busy life. From each there was a steady stream of foragers setting outward into the brilliant sunshine, and as constant a current homeward, as the bees returned heavily weighed down under loads of golden pollen from the willows by the neighbouring riverside. But round the hive, near which the bee-master presently came to a halt, there was a very different scene enacting. The deep, rich note of labour was replaced by an angry hubbub of war. The alighting-board of the hive was covered with fighting bees; company launched against company; single combats to the death; writhing masses of bees locked together and tumbling furiously to the ground in every direction. The soil about the hive was already thickly strewn with the dead and dying: and the air, for yards round, was filled with the piercing note of the fray. It seemed as hopeless to attempt to stop the carnage as it was manifestly perilous to go near.

But the bee-master had his own short way with this, as with most other difficulties. He took up a big watering-can and filled it hastily from the butt close by.

" This hive is a weak stock," he explained, " and it is being robbed by one of the stronger ones. That is always the danger in spring. We must try to drive the robbers home, and only one thing will do it. That is, a heavy rainstorm; and as there is no chance of getting the real thing, we must make one for ourselves."

He strode into the thick of the flying bees, and raising the can above his head, sent a steady cascade of water over the whole hive. The effect was instantaneous. The fighting ceased at once. The marauding bees rose on the wing and streamed away homeward. Those belonging to the attacked hive scrambled into its friendly shelter, a bedraggled, sodden crew. When at length all was quiet, the old bee-man fetched an armful of hay and heaped it up before the hive, completely covering its entire front.

" If the robbers come back," said he, " that will stop them going in, while the bees inside can crawl to and fro if they wish. But at sunset we must do away with the stock altogether by uniting it to another colony, and so put temptation out of the robbers' way. And now we must go and look for the robbers' den."

He refilled his pipe, and led the way down the long thoroughfare of the bee-city, examining every hive in turn as he passed.

" It is trouble of this kind," he said, " that does more than anything else to upset the instinct-theory of the old-fashioned naturalists, at least as far as the honey-bee is concerned. Why should a whole houseful of them suddenly break away from their old orderly industrious habits, and take to thieving and violence? But so it often happens. There is character, or the want of it, among bees just as there is in the human race. Some are gentle and others vicious; some are hard workers early and late, and others seem to take things easily, or to be subject to unaccountable moods and caprices. Then the weather has an extraordinary influence on the temper of most hives. On sunny, calm days, when the glass is ' set fair,' and the clover in full bloom, the bees will take no notice of any interference. The hives can be opened and manipulated without the slightest fear of a sting. But if the glass is falling, or the wind rising and backing, the bees will be often as spiteful as cats, and as timid as squirrels. And there are times, just before a storm, when to touch some hives would mean bringing the whole population out upon you like a nest of hornets."

He stopped by one of the hives, and laid his great sunburnt hand down flat on the entrance-board. The bees took no account of the obstacle, but ran to and fro over his fingers with perfect unconcern.

"And yet," said he, "there are bees that follow none of these general rules. Here is a stock which it is almost impossible to ruffle. You may turn their home inside out, and they will go on working just as if nothing had happened. They are famous honey-makers, while they keep to it; but, like all mild-tempered bees, they are too fond of swarming, and have to be put back into the hive two or three times before they settle down to the season's work."

As he talked, he was looking about him carefully, and at last made a short cut towards a hive standing a little apart from the rest. The bees of this hive were behaving in a very different fashion from those we had just inspected. They were running about the flight-board in an agitated way, and the whole hive gave out a note of deep unrest. The old bee-man puffed his "smoker" up into full draught, and set to work to open the hive.

"These are the honey thieves," he said, as he pulled off the coverings of the hive and laid bare its rumbling, seething interior to the searching sunlight, "and when once bees have taken to robbing their neighbours there is only one way to cure them. You must exterminate the whole brood. In the old days, a stock of bees with confirmed bad habits would be taken to the sulphur-pit and settled at once for good and all. But modern bee-keepers have a better and less wasteful way. Now, look out for the queen!"

He was lifting out the comb-frames one by one, and subjecting them to a close examination. At last, on one of the most crowded frames, he spied the huge full-bodied queen, and lifted her off by the wings. Then he closed the hive up again as expeditiously as possible.

"Now," said he, as he ground the discredited monarch under his heel, "we have stopped the mischief at the fountain-head. Of course, if we left the bees to raise another queen for themselves, she would be of the same blood as the first one, and her children would inherit the same undesirable traits. But to-morrow, when the bees are thoroughly sobered and frightened at the loss of their ruler, we will give them another full-grown fertile queen of the best blood in the apiary. In three weeks' time the new population will begin to take over the citadel; and in a month or two all the old bees will have died off, and with them the last of the robber taint."

CHAPTER XVIII

THE STORY OF THE SWARM

WHEN professional breeders of the honey-bee have succeeded in producing the much-desired non-swarming race, and swarming has become a thing of the past, naturalists of the old " instinct " school will be able to turn their backs on at least one very inconvenient question.

There is no denying that the breeders are theoretically right in their present efforts. The swarming-habit in the honey-bee is admittedly the main obstacle to large honey-takes; and now that two of the principal objects of swarming—the multiplication of stocks and renewal of queens—are fairly well understood, and can be artificially effected, there is no doubt that the universal adoption of a non-swarming strain throughout the bee-farms of the country, if such a thing were possible, would result in a very greatly increased honey-yield, and the people would get cheap honey. But at present it is not easy to see that any progress whatever in this direction has been made. The bees continue to swarm, in spite of beautifully adjusted theories; and the old attempt to fit the square peg of instinct into the round hole of fact goes on as merrily as ever.

Students of bee-life, approaching the matter unencumbered by ancient postulates, find themselves face to face with many surprising things, which would seem unexplainable on any other hypothesis than that the bees are endowed with reason, and that of no mean order.

Instinct implies invariability, a dead perfection of motive working blindly against all odds of circumstance, and always succeeding in the main. But the very essence of reason, humanly speaking, is its imperfection and continual deviation both in motive and performance. Watching a swarm of bees from the moment of its issue from the hive, the first thing that strikes the unacademic observer is that most of the bees seem to have no notion at all as to what the furore is about. They are by no means the obedient items of a common inexorable purpose. They are more like a crowd of people running in a street, all agog with excitement and curiosity, but not one of them knowing the cause of the general stampede. Sometimes a stock of bees will give visible sign of the approach of a swarming-fit for several days before the swarm actually issues. But, as often as not, no such manifestation is given. The hive, at least to the unexpert eye, seems in its normal condition right up to the moment when the great emigration takes place. And then, as at a given signal, the work suddenly stops, and the bees pour out of the hive-entrance in a living stream, darkening the air for many yards round, the cloud of darting bees rising higher and higher, and spreading over a greater space with every moment. The swarm may take three or four minutes to get fairly on the wing; and, from a populous hive, may number twenty-five or thirty thousand individuals.

There is seldom any fear of stings at such a time, and this extraordinary phase of bee-life may usually be studied at close quarters. One of the most puzzling things about it is that, however large the swarm proves to be, enough workers and drones are still left behind in the old hive to carry on the work of the stock. When the order for the sally is given, and a feverish excitement spreads at once throughout the hive, those bees chosen to remain in the old dwelling are perfectly unmoved by the general mad spirit. Directly the last of the trekking-party has gone off, the home-bees set diligently and quietly to work as if nothing had happened. With the whole garden alive with flashing wings, and resounding with the rich deep hubbub of the swarm, the bees forming the remnant of the old colony go about their usual business in perfect unconcern, lancing straight off into the sunshine towards the clover-fields, or winging busily homeward laden with honey and pollen, just as they have been doing for weeks past. And if the hive be opened at this time, it will show nothing unusual except that no queen will be found. There will be three or four queen-cells like elongated acorns hanging from the edges of the central combs; and the first queen to hatch out, and prove herself happily mated, will be allowed to destroy all the others. For the rest, work seems to be going on in a perfectly normal way. The nectar and pollen are being stored in the cells; the young grubs are being fed; most of the combs are fairly well covered with their busy population, consisting principally of young bees, although a fair sprinkling of mature workers and drones is everywhere visible. In eight or ten days the new queen will be laying and the colony rapidly regaining its former strength.

Meanwhile, the swarm is still in the air, every bee careering hither and thither with no other apparent purpose than that of allowing full vent to the mad excitement which has so mysteriously seized upon it. This state will often last a considerable time, and, in rare cases, will end by the bees trooping soberly back to the hive under just as mysterious a revulsion of feeling and resuming their old steady work. At other times the cloud of bees will suddenly rise high into the air and go straight off across country, disappearing in a few moments from the keenest view. But generally, after a short spell of this berserk frolic, the swarm seems gradually to unite under common direction. The dark network of flying bees overhead shrinks and grows denser. At last you make out the beginnings of the cluster— a mere handful of bees clinging to a branch in a tree or bush. The handful swells at a wonderful pace as the bees crowd towards it from all quarters. In three or four minutes the whole multitude is locked together in a solid pendent mass, and the wild song of freedom has died down to a few stray intermittent notes.

This silence, following the shrill, abounding turmoil, has an almost uncanny effect. It seems so utterly opposed to, and incongruous with, the mad state of things that existed before; and it is difficult to escape the conclusion that the bees have weakly

given way to an incontrollable impulse against all
their principles and inherited traditions of right, and
that now, hanging thoroughly sobered and shamed
and disillusioned, homeless and beggared, they
realise themselves face to face with the unforeseen
consequences of their thoughtless act. It is just
the conduct which might be expected of some savage
human race, pent up for long years in the rigid
bounds of an alien civilisation, which in one blind
moment has thrown to the four winds all its irksome
blessings, only to realise, when the first glowing
hour of freedom is over, that their long captivity
has made the old wild life no longer possible in fact.
Some such period of deep despondency as has come
to the silent swarm in the hedgerow can be imagined
as inevitably falling on such a race of men. But if
the conquerors were to follow the absconding tribe
into the lean wilderness and bring them home again
repentant, restoring them to their old shelter and
plenty once more, probably they would vent their
satisfaction in a chorus of joyful approval. And it
is just this which seems to be happening when the
swarm is shaken down in front of a new, well-
furnished hive. The first bees that find their way
into the cool dark interior set up a jubilant hum
unlike any other sound known in beecraft. At once
the strain is taken up by all the rest, and the whole
multitude marches into the new home to a tune
which the least fanciful must concede is nothing but
sheer satisfaction melodised.

There is little in all this which suggests a race of
creatures bound within the hard and fast laws of an
implanted instinct, which it is neither in their power
nor their pleasure to override. It is true that in the

natural life of the honey-bee this annually recurrent impulse of swarming serves several necessary ends; but the utilitarian argument, however stretched, cannot be made to explain the whole fact. There is unmistakably an element of caprice about it—a kicking over the traces—which would be natural enough in creatures possessed of reason, but totally inconceivable from any other point of view. And the farther we look into the whole problem the more perplexing it seems. If we grant that the issue of a swarm, from a hive overcrowded and headed by a queen past her prime, is a necessity, why is it that the same hive will often swarm a second and even a third time until the stock is practically extinguished and the original object of swarming wholly defeated? Or if, under the same conditions, a hive prepares to swarm and cold windy weather intervenes, how is it that frequently all idea of swarming is abandoned for the season, although apparently the necessity for it continues to exist?

Creatures which pursue a certain line of conduct under the blind promptings of instinct could hardly be credited with intelligence enough to lead them to seek another means for the desired end when the preordained means has failed. But this is just what the honey-bee appears to do in at least one instance. If the mother-bee of a colony is getting past her work, and she cannot be sent off with a swarm in the usual way, the bees will supersede her. They will deliberately put her to death, and raise another queen to take her place. This State execution of the old worn-out queens is one of the most curious and pathetic things in or out of bee-life. One probe with a sting would suffice in the matter; but the honey-bee is a great stickler for the proprieties. The royal victim must be allowed to meet her fate in a royal way; and she is killed by caresses, tight-locked in the joint embrace of the executioners until suffocation brings about her death.

CHAPTER XIX

THE MIND IN THE HIVE

STUDENTS of the ways of the honey-bee find many things to marvel at, but little to excite their wonder more than the unique system of ventilation established in the hive.

Under natural conditions it is a moot point whether bees concern themselves at all with the ventilation of their nests. Wild bees usually fix upon a site for their dwelling where there is ample space for all possible developments; and the ventilation of the home—as with most human tenements—is left pretty much to chance causes. At least, in the course of many years' observation, the writer has never seen the fanners at work in the entrance of a natural bee-settlement.

Probably this remarkable fanning system originated in a new want felt by the bees, when, in remote ages, their domestication began, and they found themselves cooped up in impervious hives which, in their very earliest form, were possibly roughly-plaited baskets, daubed over with clay, or earthen pots baked dry in the sun. This form, originally adopted by the bee-keeper as a protection against honey-thieves of all sorts, as well as against the weather, brought about a new order of things in bee-life. The free circulation of air which would obtain when the bee-colony was established naturally in a cleft of a rock or in a hollow tree became no longer possible. And so—as they have been proved to have done in many modern instances—the bees set to work to evolve new methods to meet new necessities, and the present ventilation-system gradually became an established habit of the race.

Watching a hive of bees on any hot summer's day, one very curious, not to say startling, fact must strike the most superficial observer. If the fanning bees were stationed round the flight-hole in a merely casual, irregular way, their obvious employment would be surprising enough. But it is at once seen that each fanner forms part in an ingenious and carefully thought-out plan. Outwardly, the fanners are arranged in regular rows, one behind the other, all with their heads pointed towards the hive, and all working their wings so fast that their incessant movement becomes nearly invisible. These rows of bees extend sometimes for several inches over the alighting-board, and on very hot days there may be as many as seven or eight ranks. The ventilating army never covers the whole available space. It is always at one side or the other; or, where the entrance is a wide one, it may be divided into two wings, leaving a centre space free. The fanning bees, moreover, do not keep close together, but stand in open order, so that the continual coming and going of the nectar-gatherers is in no wise impeded. There is a constant flow of worker-bees through the ranks in both directions; yet the fanning goes on uninterruptedly, and, under certain conditions, the current of air thus set up may be strong enough to blow out the flame of a candle held at the edge of the flight-board.

In all study of the ways of the honey-bee, the safer plan is to begin with the assumption that a reasoning creature is under observation, and then to work back to the surer, well-beaten tracks of thought concerning the lower creation—that is, if the observed facts warrant it. But this question of the ventilation of the modern beehive—only one of many other problems equally astounding—helps the orthodox naturalist of the old school very little on his comfortable way. We know that the wild bee generally chooses a situation for her nest which is neither cramped nor confined, but has in most cases ample space available for the future growth of the colony. Security from storm or flood seems to be the first consideration. The fact that the interior of a bee-nest is more or less in darkness appears to be mainly accidental. Bees have no particular liking for absolute darkness, nor, in fact, is any hive perfectly free from light. Experiment will prove that a very small aperture is sufficient to admit a considerable amount of reflected and diffused light, quite enough for the needs of the hive. It may be supposed, therefore, that the bees would have no objection to building in broad daylight, or even sunlight, if, in conjunction with the first necessities of shelter, security, and equable temperature, such a location were easily obtainable under natural conditions It would only be another instance of their unique adaptability to circumstances forced upon them.

In the matter of ventilation, however, they seem to make a very determined and highly successful stand against imposed conditions. Bee-keeping cannot be made a profitable occupation unless the work of the bees is kept strictly within certain sharply-defined limits, and probably the modern movable comb hive is the best means to this end. That it leaves the necessity of ventilation wholly unprovided for is not the fault of the bee-master, but of the bees themselves. They refuse point-blank to have anything to do with human notions of hygiene. Many devices have been tried, in the form of vent-shafts and the like, to carry off the vitiated air of the hive, but all have failed, because the bees insist on stopping up every crack or crevice left in walls, roof, or floor. For some inscrutable reason they will have only the one opening, which must serve for all purposes, and the hive-maker has had to learn by hard-won experience that the bees are right.

Perhaps, in any attempt to follow the reasoning of the bees in this matter, it is well first of all to get rid of the word " fanning " altogether. The wing-action of the ventilating bees is more that of a screw-propeller than a fan. The air is not beaten to and fro, as a fan would beat it, but is driven backwards, and thus the ventilating squadron on the flight-board really sets up an exhaust-current, which draws the contaminated air out of the hive. This implies an equally strong current of fresh air passing into the hive, and explains why the bees work at the side of the entrance only, the central, unoccupied space being obviously the course of the intake. Thus the bees' system of ventilation can be described as a swiftly-flowing loop of air, having both extremities outside the hive, much as a rope moves over a pulley, and it can be readily understood that any supplementary inlet or outlet—such as the bee-master would instal, if he were permitted —would be rather a hindrance to the system than a help. Probably the actual main current keeps to the walls of the hive throughout, the ventilation between the brood-combs being more slowly effected. This would fulfil a double purpose. The air supplied to the central portion, or brood-nest proper, would be thoroughly warmed before it reached the young larvæ, while the outer and upper combs, where the stores of new honey are maturing, would lie in the full stream.

It must be remembered that a constant supply of fresh air of the right temperature is as necessary for the brewing honey as it is for the bees and young brood. The nectar, as gathered from the flowers, needs to be deprived of the greater part of its moisture before it becomes honey. Thus, in the course of the season, many gallons of water must pass out of the hive in the form of vapour, and the removal of this water constitutes an important part of the work of the ventilating army. Here, again, the wisdom of the bees in insisting on a mechanical, as opposed to an automatic, system of air-renewal, becomes evident. If the warm, moisture-laden air were left to discharge itself from the hive by its own buoyancy, condensation of this moisture would take place on the cooler surfaces of the hive-walls, and the lower regions of the hive would speedily become a quagmire. But by setting up a mechanically-driven current the air is drawn out before condensation can take place, and thus, in one operation, forming a veritable triumph in economics, the hive interior is rendered both dry and salutary, while its temperature is sustained at the necessary hatching-point for the young brood.

A reflection which will occur to most thinking minds is, why should the domesticated honey-bee be constrained to resort to all these devices, when the wild bee seems to lead a happy-go-lucky existence, comparatively free, so far as we know, from such complicated cares? The answer to this is that the science of apiculture has wrought a change in the bees' normal environment which is probably without parallel in the whole history of the domestication of the lower creatures. In a modern hive the honey-bee lives on a vastly elaborated scale, and the ancient rules of bee-life are no longer applicable. Much the same sort of thing has happened as in the case of a village which has grown to a city. It is useless to deal with the new order of things as a mere question of arithmetic. Abnormal growth in a community involves change not only in scale but in principle; and it is the same with a hive of bees as with a hive of men.

CHAPTER XX

THE KING'S BEE-MASTER

STUDENTS of old books on the honey-bee—and perhaps there has been more written about bees during the last two thousand years than of all other creatures put together—do not quite know what to make of Moses Rusden, who was Charles the Second's bee-master, and wrote his " Further Discovery of Bees " in the year 1679. The wonder about Rusden is that obviously he knew so much that was true about bee-life, and yet seems, of set purpose, to have imparted so little. He was a shrewdly observant man, of lifelong experience in his craft. His system of bee-keeping would not have disgraced many an apiculturist of the present time, often yielding him a honey harvest averaging sixty pounds to the hive, which is a result not always achieved even by our foremost apiarian scientists. His hives were fitted with glass windows, through which he was continually studying his bees. He must have had endless opportunities of proving the fallacy and folly of the ancient classic notions as to bee-life. And yet we find him gravely upholding almost the entire framework of fantastic error, old even in Pliny's time; and speaking of the king-bee with his generals, captains, and retinue, honey that was a dew divinely sent down from heaven, the miraculous propagation of bee-kind from the flowers, and all the other curious myths and fables handed down from writer to writer since the very earliest days.

But, reading on in the little time-stained, worm-eaten book, it is not very difficult to guess at last why Rusden adopted this attitude. He was the King's bee-master, and therefore a courtier first and a naturalist afterwards. In the first flush of the Restoration, anyone who had anything to say in support of the divine right of kings was certain to catch the Royal eye. Rusden admits himself conversant with Butler's " Feminine Monarchie," published some fifty years before, in which the writer argues that the single great bee in a hive was really a female. To a man of Rusden's practical experience and deductive quality of mind, this statement must have lead, and no doubt did lead, to all sorts of speculations and discoveries. But with a ruler of Charles the Second's temperament, feminine monarchies were not to be thought of. Rusden saw at once his restrictions and his peculiar opportunity, and wrote his book on bees, which is really an ingenious attempt to show that the system of a self-ruling commonwealth is a violation of nature, and that, whether for bees or men, government under a king is the divinely ordained state.

Whether, however, Rusden was deliberately insincere, or actually succeeded in blinding himself conveniently for his own purposes, it must be admitted not only that he argued the case with singular adroitness, but that never did facts adapt themselves so readily to either conscious or unconscious misrepresentation. In the glass-windowed hives of the Royal bee-house at Saint James's, he was able to show the King a nation of creatures evidently united under a common rule, labouring together in harmony and producing works little short of miraculous to the mediæval eye. He saw that these creatures were of two sorts, each going about its duty after its kind, but that in each colony there was one bee, and only one, which differed entirely from the rest. To this single large bee all the others paid the greatest deference. It was cared for and nourished, and attended assiduously in its progress over the combs. All the humanly approved tokens of royalty were manifest about it. No wonder the King's bee-master was not slow in recognising that, in those troublous times, he could do his patron no greater service than by pointing out to the superstitious and ignorant multitude—still looking askance at the restored monarchy—such indisputable evidence in nature of Charles's parallel right.

And perhaps nature has never been at such pains to conceal her true processes from the vulgar eye as in this case of the honey-bee. If Rusden ever suspected that the one large bee in each colony was really the mother of all the rest, and had set himself to prove it, he would have found the whole array of visible facts in opposition to him. If ever a truth seemed established beyond all reasonable doubt, it was that the ordinary male-and-female principle, pertaining throughout the rest of creation, was abrogated in the single instance of the honey-bee. The ancients explained this anomaly as a special gift from the gods, and the bees were supposed to discover the germs of bee-life in certain kinds of flowers and to bring them home to the cells for development. Rusden improved upon this idea by assigning to his king-bee the duty of fertilising these embryos when they were placed in the cells, for he could not otherwise explain a fact of which he was perfectly well aware—that the large bee travelled the combs unceasingly, thrusting its body into each cell in turn. Rusden also held that the worker-bees were females, but only—as Freemasons would say—in a speculative manner. They neither laid eggs nor bore young. Their maternal duties consisted only in gathering the essence of bee-life from the blossoms and nursing and tending the young bees when they emerged from their cradle-cells. The drones were a great difficulty to Rusden. To admit them to be males—as some held even in his day—would have been against the declared object of his book, as tending to entrench upon royal prerogatives. Luckily, this truth was as easy of apparent refutation as all the rest. No one had ever detected any traffic of the sexes amongst bees either in or out of the hives; nor, indeed, is such detection possible. The fact that the queen-bee has concourse with the drone only once in her whole life, and that their meeting takes place in the upper air far out of reach of human observation, is knowledge only of yesterday. In Rusden's time such a marvel was never even suspected. As the drones, therefore, were never seen to approach the worker bees or to notice them in any way, and as also young bees were bred in the hives during many months when no drones existed at all, Rusden's ingenuity was equal to the task of bringing them into line with his theory.

If he had lived a few decades earlier, and it had been Cromwell, instead of the heartless, middle-aged rake of a sovereign, whom he had to propitiate, no doubt Rusden would have asked his public to swallow Pliny's whole apiarian philosophy at a gulp. Bee-life would then have been held up as a foreshadowing of celestial conditions, and the facts would have lent themselves to this view equally as well. But his task was to represent the economy of the hive as a clear proof of divine authority in kingship, and it must be conceded that, as far as knowledge went in those days, he established his case.

His book was published under the ægis of the Royal Society, and " by his Majestie's especial Command," which was less a testimony of the King's love for natural history than of his political astuteness. Apart, however, from its peculiar mission, the book is interesting as a sidelight on the old bee-masters and their ways. Probably it represents very fairly the extent of knowledge at the time, which had evidently advanced very little since the days of Virgil. Rusden taught, with the ancients, that honey was a secretion from the stars, and that wax was gathered from the flowers, as well as the generative matter before mentioned. He had one theory which seems to have been essentially his own. The little lumps of many-coloured pollen, which the worker-bees fetch home so industriously in the breeding season, he held to be the actual substance of the young bees to come, in an elementary state. These, he tells us, were placed in the cells, having absorbed the feminine virtues from their bearers on the way. The king-bee then visited each in turn, vivifying them with his essence, after which they had nothing to do but grow into perfect bees. He got over the difficulty of the varying sexes of the bees bred in a hive by asserting that these lumps of animable matter were created in the flowers, either female, or neuter—as he called the drones—or royal, as the case might be. Having denied the drones any part in the production of their species, or in furnishing the needs of the hive, Rusden was hard put to it to find a use for them in a system where it would have been *lèse-majesté* to suppose anything superfluous or amiss. He therefore hits upon an idea which, curiously enough, embodies matter still under dispute at the present time, although it is being slowly recognised as a truth. Rusden says the use of the drones is to take the place of the other bees in the hive when these are mostly away honey-gathering. Their great bodies act as so many warming stoves, supplying the necessary heat to the hatching embryos and the maturing stores of honey. It is well known that drones gather together side by side, principally in the remoter parts of the hive, often completely covering these outer combs. They seldom rouse from their lethargy of repletion to take their daily flight until about midday, when most of the ingathering work is over, and the hive is again fairly populous with worker-bees. Probably, therefore, Rusden was quite right in his theory, which, hundreds of years after, is only just beginning to be accepted as a fact.

CHAPTER XXI

POLLEN AND THE BEE

POPULAR beliefs as to the ways of the honey-bee, unlike those relating to many other insects, are surprisingly accurate, so far as they go. But, dealing with such a complex thing as hive-life, it is well-nigh impossible to have understanding on any single point without going very much farther than the ordinary tabloid-method of knowledge can carry us. This is especially true with regard to pollen, and the uses to which it is put within the hive. The hand-books on bee-keeping usually tell us that pollen is employed with honey as food for the young bees when in the larval state; but this is so wide a generalisation that it amounts to almost positive error. As a matter of fact, the pollen in its raw condition is given only to the drone-larva, and this only towards the end of its life as a grub. For the first three days of the drone-larva's existence, and in the case of the young worker-bee for the whole five days of the larval period, the pollen is administered by the nurse-bees in a pre-digested state. After partial assimilation, both the pollen and the nectar are regurgitated by these nurse-bees, and form together a pearly-white fluid—veritable bee-milk—on which the young grubs thrive in an extraordinary way.

There are few things more fascinating than to watch a hive of bees at work on a fine June morning, and to note how the pollen is carried in. With a prosperous stock, thousands of bees must pass within the space of a few minutes, each bee dragging behind her a double load of this substance. Very often, in addition to the half-globes of pollen which she carries on her thighs, the bee will be smothered in it from head to foot, as in gold-dust. If you track her into the hive, one curious point will be noted. No matter how fast she may go, or what frantic spirit of labour may possess the entire colony, the pollen-laden bee is never in a hurry to get rid of her load. She will waste precious time wandering over the crowded combs, continually shaking herself, as though showing off her finery to her admiring relatives; and it may be some minutes before she finally selects a half-filled pollen-cell and proceeds to kick off her load. The different kinds of pollen are packed into the cells indiscriminately, the bee using her head as a ram to press each pellet home. When the cell is full it is never sealed over with a waxen capping, as in the case of the honey-stores, but is left open or covered with a thin film of honey, apparently to preserve it from the air. The nurse-bees, who are the young workers under a fortnight old, help themselves from these pollen-bins. They also frequently stop a pollen-bearer as she hurries through the crowd, and nibble the pollen from her thighs.

Throughout the season there is hardly an

imaginable colour or shade of colour which is not represented in the pollen carried into a beehive; and with the aid of a microscope it is not difficult to identify the source of each kind. In May, before the great field-crops have come into bloom, the pollen is almost entirely gathered from wild flowers, and consists of various rich shades of yellow and brown. By far the heaviest burdens at this time are obtained from the dandelion. The pollen from this flower is a peculiarly bright orange, and is easily recognised under a strong glass by its grains, which are in the form of regular dodecahedrons, thickly covered all over with short spikes.

It is well known that the honey-bee confines herself during each journey to one species of flower, and this is proved by the microscope. It is not easy to intercept a homing bee laden with pollen. On alighting before the hive she runs in so quickly that the keenest eye and deftest hand are necessary to effect her capture. But with the aid of a miniature butterfly-net and a little practice it can generally be done; and then the pellet of pollen will be found to consist almost invariably of one kind of grain. But it is not always so. The honey-bee, as a reasoning creature, does not and cannot be expected to do anything invariably. Among some hundreds of these pollen-lumps examined under the microscope I have occasionally found grains of pollen differing from the bulk. Perhaps there are no two species of flower which have pollen-grains exactly alike in colour, shape, and size, and in most the differences are very striking. In the cases mentioned the bulk of the pollen was made up of long oval yellow grains divided lengthwise into

three lobes or gores, which were easily identifiable
as coming from the figwort. The isolated grains
were very minute spheres thickly studded with
blunt spikes—obviously from the daisy. The figwort
is a famous source of bee-provender in spring
time, and its pollen can be seen flowing into the
hives at that time in an almost unbroken stream of
brilliant chrome-yellow. The brownish-gold masses
that are also being constantly carried in are from
the willow; and where the hives are near woodlands the bluebells yield the bees enormous
quantities of pollen of a dull yellowish white.

It is interesting that all these various materials,
so carefully kept asunder when gathered, are for
the most part inextricably mingled within the hive.
Obviously the system of visiting only one species
of flower on each foraging journey can have no
relation to pollen-gathering; nor does it seem to
apply to the nectar obtained at the same time. It
cannot be inferred that the contents of each honey-
cell are brewed from only one source, because it has
been proved that bees do blend the various nectars
together when several crops are simultaneously in
flower. A honey-judge can easily detect the flavours
of heather and white-clover in the same sample of
honey by taste alone. But there is another and
much more conclusive way of deciding the source
from which a particular sample of honey has been
obtained. In the purest and most mature honeys
there are always a few accidental grains of pollen,
invisible to the eye, yet easily detected under a
strong glass. And these may be taken as almost
infallible guides to the species of flowers visited by
the foraging bees. The only explanation which

seems possible, therefore, of the honey-bee's care to visit only one kind of blossom on each journey is that it is done for the sake of the plant itself, cross-fertilisation being thus rendered extremely improbable.

When once the bee-man has succumbed to the fascination of the microscope, there is very little chance that he will ever return to his old panoramic view of things. He goes on from wonder to wonder, and the horizon of the new world he has entered continually broadens with each marvelling step. To the old rule-of-thumb bee-keepers pollen was mere bee-bread; and the fact that the bees preferred one kind to another did not greatly concern them. But at a time when the small-holder is beginning to feel his feet, and the question of the feasibility of planting for bee-forage is certain to arise, it is necessary to know why bees gather this important part of their diet from particular kinds of flowers, while leaving severely alone others which appear to be equally attractive. To this question the microscope supplies a sufficient answer.

Chemists have determined that nectar is the heat and force-producer in the food of the bee, while pollen supplies its nitrogenous tissue-building qualities. It is evident that bees select certain pollens for their superior nutritive powers, just as in bread-making we prefer wheat to any other species of grain. In the kinds of pollen most in favour with bees a good microscope will reveal the fact that the pollen-grains are often accompanied by a certain amount of true farina, as well as essential oils, which must greatly enhance their food-value. And in those crops generally neglected by bees, such as daisies and buttercups, those accompaniments appear to be absent. The dandelion is especially rich in a thick yellow oil, which the bees carry away with the pollen; while two plants in particular of which the bees are especially fond—the crocus and the box—have a large amount of this farina mingled with the true pollen.

It is only within the last century or so that the real uses of pollen in the economy of the hive have been ascertained. Until comparatively recent times the pollen was supposed to be crude wax, which the bees refined and purified into the white ductile material of the new combs; and a few old-fashioned bee-keepers still hold this view, and refuse to believe that the wax used in comb-building is entirely a secretion from the bee's own body. Pollen, indeed, seems to have very little to do with wax, hardly any nitrogenous food being consumed while the wax is being generated.

CHAPTER XXII

THE HONEY-FLOW

ON Warrilow Bee-Farm, where it lay under the green lip of the Sussex Downs, there was always food for wonder, whether the year was at its ebb or its flow. But in July of a good season the busy life of the farm reached a culminating point.

The ordinary man, in search of excitement, distraction, the heady wine served out only to those who stand in the fighting-line of the world, would hardly seek these things in a little sleepy village sunk fathoms deep in English summer greenery. But, nevertheless, with the coming of the great honey-flow to Warrilow came all these subtle human necessities. If you would keep up with the bee-master and his men at this stirring time, you must be ready for a break-neck gallop from dawn to dusk of the working day, and often a working night to follow. While the honey-flow endured, muscles and nerves were tried to their breaking-point. It was a race between the great centrifugal honey-extractor and the toiling millions of the hives; and time and again, in exceptionally favourable seasons, the bees

158

would win; the honey-chambers would clog with the interminable sweets, and the dreaded atrophy of contentment would seize upon the best of the hives, with the result that they would gather no more honey.

A week of hot bright days and warm still nights, with here and there a gentle shower to hearten the fields of clover and sainfoin; and then the fight between the bee-master and his millions would begin in earnest. There would be no more quiet pipes, strolling and talking among the hives: the Bee-Master of Warrilow was a general now, with all a great commander's stern absorption in the conduct of a difficult campaign. Often, with the first grey of the summer's morning, you would hear his footsteps on the red-tiled path of the garden below, as he hurried off to the bee-farm, and presently the bell in the little turret over the extracting-house would clang out a reveille to his men, and draw them from their beds in the neighbouring village to another day of work, perhaps the most trying work by which men win their bread.

It is nothing in the ordinary way to lift a super-chamber weighing twenty pounds or so. But to lift it by imperceptible degrees, place an empty rack in its place, return the full rack to the hive as an upper story, and to do it all so quietly and gently that the bees have not realised the onslaught on their home until the operation is complete, is quite another thing. And a long day of this wary, delicate handling of heavy weights, at arm's length, under broiling sunshine, is one of the most nerve-wearing and back-breaking experiences in the world.

One of the mistakes made by the unknowing in bee-craft is that the bee-veil is never used among professional men. But the truth is that even the oldest, most experienced hand is glad enough, at times, to fall back behind this, his last line of defence. All depends upon the momentary temper of the bees. There are times when every hive on the farm is as gentle as a flock of sheep, and it is possible to take any liberty with them. At other times, and apparently under much the same conditions, stocks of bees with the steadiest of reputations will resent the slightest interference, while the mere approach to others may mean a furious attack. No true bee-man is afraid of the wickedest bees that ever flew, but it is only the novice who will disdain necessary precautions. Even the Bee-Master of Warrilow was seldom seen without a wisp of black net round the crown of his ancient hat, ready to be let down at a moment's notice if the bees showed any inclination to sting.

In a long vista of memorable days spent at Warrilow, one stands out clear above all the rest. It was in July of a famous honey-year. The hay had long been carried, and the second crops of sainfoin and Dutch clover were making their bravest show of blossom in the fields. It was a stifling day of naked light and heat, with a fierce wind abroad hotter even than the sunshine. The deep blue of the sky came right down to the earth-line. The farthest hills were hard and bright under the universal glare. And on the bee-farm, as I came through the gap in the dusty hedgerow, I saw that every man had his veil close drawn down. The bee-master hailed me from his crowded corner.

" Y'are just to the nick!" he called, in his broadest Sussex. " 'Tis stripping-day wi' us, an' I can do wi' a dozen o' ye! Get on your veil, d'rectly-minute, an' wire in t'ot!"

The fierce hot wind surged through the little city of hives, scattering the bees like chaff in all directions, and rousing in them a wild-cat fury. Overhead the sunny air was full of bees, striving out and home; and from every hive there came a shrill note, a tremulous, high-pitched roar of work, half-baffled, driven through against all odds and hindrances, a note that bore in upon you an irresistible sense of fear. I pulled on the bee-veil without more ado.

" Stripping-day " was always the hardest day of the year at Warrilow. It meant that some infallible sign of the approaching end of the harvest had been observed, and that all extractable honey must be immediately removed from the hives. A change of weather was brewing, as the nearness of the hills foretold. There might be weeks of flood and tempest coming, when the hives could not be opened. Overnight there had been a ringed moon, and the morning broke hot and boisterous, with an ominous clearness everywhere. By midday the glass was tumbling down. The bee-master took one look at it, then called all hands together. "Strip!" he said laconically; and all work in extracting-house and packing-sheds was abandoned, and every man braced himself to the job.

The hives were arranged in long double rows, back to back, with a footway between wide enough to allow the passage of the honey barrow. This was not unlike a baker's hand-cart, and contained empty combs, which were to be exchanged for the full combs from the hives. I found myself sharing a row with the bee-master, and already infused with the glowing, static energy for which he was renowned. The process of stripping the hives varied little with each colony, but the bees themselves furnished variety enough and to spare. In working for comb-honey, the racks or sections are tiered up one above the other until as many as five stories may be built over a good stock. But where the honey is to be extracted from the comb another system is followed. There is then only one super-chamber, holding ten frames side by side, and these frames are removed separately as fast as the bees fill and seal them, their place being taken by the empty combs extracted the day before.

The whole art of this work consists in disturbing the bees as little as possible. At ordinary times the roof of the hive is removed, the " quilts " which cover the comb-frames are then very gently peeled away, and the frames with their adhering bees are placed side by side in the clearing-box. The honey-chamber is then furnished with empty combs, and the coverings and roof replaced. On nine days out of ten this can be done without a veil or any subduing contrivance; and the bees which were shut up with the honey in the clearing-box will soon come out through the traps in the lid and fly back to their hives. But when time presses, and several hundred hives must be gone through in a few hours, a different system is adopted. Speed is now a main desideratum in the work, and on stripping-day at Warrilow resort is made to a contrivance seldom seen there at other times. This is simply a square of cloth saturated with weak carbolic acid, the most detested, loathsome thing in bee-comity. Directly the comb-frames are laid bare these cloths are drawn over them, and in a few moments every bee has crowded down terror-stricken into the lower regions of the hive, leaving the honey-chamber free for instant and swift manipulation.

CHAPTER XXIII

SUMMER LIFE IN A BEE-HIVE

IF you go to the bee-garden early of a fine summer's morning you will be struck by the singular quiet of the place. All the woods and hedgerows are ringing with busy life. The rooks are cawing homeward with already hours of strenuous work behind them. The cattle in the meadows are well through their first cud. But as yet the bee-city is as still as the sleeping village around it. Now and again a bee drops down from the sky on a deserted hive-threshold with sleepy hum, and runs past the guards at the gate. But these are bees that have wandered too far afield overnight, tempted by the sunny warmth of the evening. The dusk has caught them, and obliterated their flying-marks. They have perforce camped out under some broad leaf, to be wakened by the earliest light of morning and hurry home with their belated loads.

The sun is well up over the hillbrow before the visible life of the bee-garden begins to rouse in earnest. The water-seekers are the first to appear.

Every hive has its traditional dipping-place, generally the oozy margin of some neighbouring pond, where the house-martins have been wheeling and crying since the first grey of dawn. Now the bees' clear undertone begins to mingle with the chippering chorus. In a little while there is a thin straight line of humming music stretched between the hives and the pond: it could not be straighter if a surveyor had made it with his level. Again a little while, and this long searchlight of melody thrown out by the bee-garden veers to the north. You may track it straight over copse and meadow, seeing not a bee overhead, but guided unerringly by the arrow-flight of music, until, on the far hillside, it is lost in a perfect roar of sound. Here the white-clover is in almost full blossom again: in southern England at least it is always the second crop of clover that yields the most plentiful harvest to the hives.

It must be a disturbing thing to those kindergarten moralists who hold the bee up to youth for an example of industry and prudence to learn that she is by no means an early riser; though, at this time of year, she is undoubtedly both wealthy and wise. For it is her very wisdom that now makes her a lie-abed. When the iron is hot, she will not be slow in striking. But it is nectar, not dewdrops, from which she makes her honey. Very wisely she waits until the sun has drunk up the dew from the clover-bells, and then she hurries forth to garner their undiluted sweets. Even then, perhaps, three-fourths of her burden will be carried uselessly. In the brewing-vats of the hive the nectar must stand and steam until three parts of its original bulk has evaporated, and its sugar has been inverted into grape-sugar. Then it is honey, but not before. When we see the fanning-army at work by the entrance of a hive, it is not alone an undoubted passion for pure air that moves the bees to such ingenious activity. In the height of the honey season many pints of vaporised liquid must be given off by the maturing stores in the course of a day and night, and all this water must be got rid of. Herein is shown the wisdom of the bee-master who makes the walls of his hives of a material that is a bad conductor of heat. It is a first necessity of health to the bees that the moisture in the air, which they are incessantly fanning out at this time, should not condense until it is safely wafted from the hive. A cold-walled hive can easily become a quagmire.

The bee-garden is quiet now in the sweet virgin light of the summer's morning; but the thought of it as containing so many houses of sleep, true of the village with its thatched human dwellings, could not well be farther from the truth in regard to the village of hives. There is little sleep in a bee-hive in summer. Of any common period of rest, of any quiet night when all but the sentinels at the gate are slumbering, of any general time of relaxation, there is absolutely none. Each individual bee—forager or nurse, comb-builder or storekeeper—works until she can work no more, and then stops by the way, or crawls into the nearest empty cell for a brief siesta. But the life of the hive itself never halts, never wavers in summer-time, night or day. Go to it morning, noon, or night in the hot July season, and you will always find it driving onward unremittingly. The crowd is surging to and fro. There is ever the busy deep labour-note. Its people are building, brewing, wax-making, scavenging, wet-nursing, being born and dying: it is all going on without pause or break inside those four reverberating walls, while you stand without in the dew-soaked grass and level sunbeams wondering how it is that all the world can be at full flood-tide of merry life and music while these mysterious hive people give scarce a sign.

It is at night chiefly that the combs are built. The wax, that is a secretion from the bees' own bodies, will generate only under great heat, and the temperature of the hive is naturally greatest when all the family is at home. In the night also such works as transferring a large mass of honey from one comb to another are undertaken. It is curious to note that at night time the drones get together in the remotest parts of the hive, apparently to keep up the heat in these distant quarters, which are away from the main cluster of worker-bees. There is hardly another thing in creation, perhaps, with a worse name than the drone-bee. But like all bad things he is not so bad as he is represented. Apart from his main and obvious use, the drone fulfils at least one very important office. His habit is not to leave his snug corner until close upon midday. Thus, when every able-bodied worker bee is out foraging, the temperature of the hive is sustained by the presence of the drones, and the young bee-brood is in no danger of chilling.

Though the supreme direction of all affairs in a

bee-hive falls to the lot of the worker-bees, the
queen-mother is second to none in industry. At this
time of year she goes about her task with a dogged
patience and assiduity pathetic to witness. She may
have to supply from two thousand to three thousand
brood-cells with eggs in the course of a single day,
and she is for ever wandering through the crowded
corridors of the hive looking for empty cradles.
The old bee-masters believed that the queen was
always accompanied in these unending promenades
by exactly a dozen bees, whom they called the
Twelve Apostles. It is true that whenever the
queen stops in her march she is immediately
surrounded by a number of bees, who form themselves into a ring, keeping their heads ceremoniously
towards her. But close observation reveals the
fact that the queen-bee is never followed about by a
permanent retinue. When she moves to go on, the
ring breaks and disperses before her; but the bees
who gather round her on her next halt are those who
happen to occupy the space of comb she has then
reached.

The truth seems to be that she is passed from
" hand to hand " over the combs of the brood-nest,
and is stopped wherever a cell requires replenishing.
Each bee that she encounters on her path turns
front and touches her gently with her antennæ.
The queen constantly returns these salutes as she
moves, and it looks exactly as if she were going the
rounds of her domain and collecting information.
Often she is stopped by half a dozen bees in a solid
phalanx, and carefully headed off in a new direction.
She looks into every cell as she goes, and when she
has lowered her body into a cell, the Apostles

instantly gather about her, with strokings and caresses. But their number is seldom twelve. It varies according to the bulk and length of the queen herself, and is more often sixteen than a dozen.

CHAPTER XXIV

THE YELLOW PERIL IN HIVELAND

IN the hedgerow that surrounds the bee-garden the wrens and robins have been singing all the morning long. Still a few pale sulphur buds remain on the evening-primroses. The balsams make a glowing patch of majenta by the garden gate. Over the door porch of the old thatched cottage purple clematis climbs bravely; and the nasturtiums still flaunt their scarlet and gold in the sunny angle of the wall. But, for all the colour and the music, the hot sun, and the serene blue air overhead, you can never forget that it is October. If the towering elm-trees by the lane-side showed no fretting of amber in their greenery, nor the beeches sent down their steady rain of russet, there would still be one indubitable mark of the season—the voice of the hives themselves.

Rich and wavering and low in the sweet autumn sunlight, it comes over to you now with the very spirit of rest in every halting tone. There is work, of a kind, doing in the bee-garden. A steady tide of bees is stemming out from and home to every hive. But there is none of the press and busy clamour of bygone summer days. It is only a make-believe of duty. Each bee, as she swings up into the sunshine, hovers a while before setting easy sail for the ivy in the lane; and, on returning, she may bask for whole minutes together on the hot hive-roof. There is no sort of hurry; little as there may be to do abroad, there is less at home.

But to one section of the bee-community, these slack October hours bring no cessation of toil. The guards at the gate must redouble their vigilance. Cut off from most of their natural supplies, the yellow pirates—the wasps—are continually prowling about the entrance; and, in these lean times, will dare all dangers for a fill of honey. Incessant fierce skirmishes take place on the alighting-board. The guards hurl themselves at each adventuress in turn. The wasp, calculating coward that she is, invariably declines battle, and makes off; but only to return a little later, hoping for the unwary moment that is sure to come. While the whole strength of the picket is engaged with other would-be pilferers, she slips round the scuffling crew, and plunges into the fragrant gloom of the hive.

The variation in temperament among the members of a bee-colony is never better illustrated than by the way in which these marauders are received and dealt with. The wasp never tries to pick a way to the honey-stores through the close packed ranks of the bees. She keeps to the sides of the hive, and works her way up by a series of quick darts whenever a path opens before her. Evidently her plan is to avoid contact with the home-keeping bees, which, at this time of year, have little more to do than loiter over the combs, or tuck themselves away in the empty brood-cells by the hour together. But in her desultory advance, she often cannons against single bees; and then she may be either mildly interrogated, fiercely challenged, or may be allowed to pass with a friendly stroke of the antennæ, as though she were an orthodox member of the hive. Again, you may see her recognised for a stranger by three or four workers simultaneously. She will be surrounded and closely questioned. The bees draw back and confer among themselves in obvious doubt. The wasp knows better than to await the result of their deliberations; by the time they look for her again, she is gone.

She carries her life in her hand, and well she knows it. The farther she goes, the more suspicious and menacing the bees become. Now she has wild little scuffles here and there with the boldest of them, but her superior adroitness and pace save her at every turn. It is about an even wager that she will reach the brimming honey-cells, load herself up to the chin, and escape home to her paper-stronghold with her spoils.

As often as not, however, these hive-robbing wasps pay the last great price for their temerity. Those who study bee-life closely and unremittingly, year after year, find it difficult to escape the conclusion that there are certain bees in the crowd who are mentally and physically in advance of their sisters. The notion of the old bee-keepers—that there were generals and captains as well as rank-and-file in the hive—seems, in fact, to be not entirely without latter-day confirmation. And it is just the chance of falling in with one of these bees that constitutes, for the wasp, the main risk when robbing the hives.

If this happens, there is no longer any doubt of the turn affairs are to take. At an unlucky moment the wasp brushes against one of these hive-constables and instead of indifference, or, at most, a spiteful tweak of the leg or wing in passing, she finds herself suddenly at deadly grips. The bee's attack is as swift as it is furious. Seizing the yellow honey-thief with all six legs, she hacks away at her with her jaws, at the same time curving her body inwards with her cruel sting bared to the hilt. Even now, although more than equal to one bee at any time, the policy of the wasp is to refuse the fight, and to run. Her long legs give her a better reach. She forces her adversary away, disengages, and charges off towards the dim light of the entrance.

In all that follows, this is the beacon that guides her. If she could get a clear course, her greater speed would soon out-distance all pursuit. But the sudden clash of arms in the quiet of the hive has an extraordinary effect on the sluggish colony. The alarm spreads on every side. Wherever the wasp runs now she is met with snapping jaws and detaining embraces. As she rushes madly down the comb, she is continually pulled up in full flight by bees hanging on to her legs, her wings, her black waving antennæ. A dozen times she shakes them all off, and speeds on, the spot of light and safety in the distance ever growing brighter and larger. But she seldom escapes with her life if affairs have reached this pass. The way now is alive with enemies. She is stopped and headed off in all directions. Trying this way and that for a loophole, she finally gives it up and turns on her tracks, bewildered and panic-stricken, only to rush straight into the midst of more foes.

The end is always the same. Another of the stalwarts spies her, and in a moment the two are locked in berserk conflict. Together they drop down between the combs and thud to the bottom of the hive. Here it is hard to tell what happens. The fight is so fierce and sharp, and the two whirl round and tumble over and over together so wildly that you can make out little else than a spinning blur of brown and yellow. A great bright drop of honey flies off: in her extremity the wasp has disgorged her spoils. Perhaps for an instant the warriors may get wedged up in a corner, and then you may see that they are not lunging at random with their stilettos, but each is trying for a side-thrust on the body; these mail-clad creatures are vulnerable to each other only at one point—the spiracles, or breathing-holes. Often the wasp deals the first fatal blow, and the bee drops off mortally hurt. She may even dispose of three or four of her assailants thus in quick succession. But each time another bee closes with her at once. For the wasp there can only be one end to it. Sooner or later she gets the finishing stroke.

And then there follows a grim little comedy. The bee, torn and ragged as she is from the incessant gnashing of those razor-edged yellow jaws, nevertheless pauses not a moment. She grips her dying adversary by the base of the wing, and struggles off with her towards the entrance of the hive. It is a hard job, but she succeeds at last. Alternately pushing her burden before her, or dragging it behind, at length she wins out into the open, and, with a final desperate effort, tumbles the wasp over the edge of the footboard down into the grass below. Yet this is not enough. The victory must be celebrated in the old warrior fashion. Rent and bleeding and exhausted as she is, she finds she can still fly. And up into the mellow sunbeams of the October morning she sweeps, giddily and uncertainly, piercing the air with her shrill song of triumph. Through the murmurous quiet of the bee-garden, it rings out like a cry in the night.

CHAPTER XXV

THE UNBUSY BEE

IT is well-nigh two months now since the hives were packed down for the winter, and the bees are flying as thick as on many a summer's day.

Yet no one could mistake their flight for the summer flight. It is not the straight-away eager rush up into the blue vault of the sunny morning—high away over hedgerow and village roof-top towards the clover-fields, whitening the far-off hillside with their tens of thousands of honey-brimming bells. It is rather the vagrant, purpose-less hanging-about of an habitually busy people forced to make holiday. Through it all there runs the pathetic interest in trifles, half-hearted and wholly artificial, that you see among the lolling crowd of men when a great strike is on—the thoughtful kicking at odd pebbles; stride-measuring on the flag-stones; little vortices of excitement got up over minute incidents that would otherwise pass unnoticed; the earnest flagellation of memory over past happenings more trivial still.

Thus the bees idle about and wander, on this

176

still November morning, doing just the things you would never expect a bee to do. The greater number of them merely take long desultory reaches a-wing through the sunshine, going off in one objectless direction, turning about at the end of a few yards with just as little apparent reason, coming back to the hive at length on no more obvious errand than that, where there is nothing to do, doing it in another place bears at least the semblance of achievement.

But many of them succeed in conjuring up an almost ludicrous assumption of business. One comes driving out of the hive-entrance at a great pace, designedly, as you would think, going out of her way to bustle the few bees lounging there, as if the entrance-board were still thronged with the streaming crowd of summer days foregone. She stops an instant to rub her eyes clear of the hive-darkness; tries her wings a little to make sure of their powers for a heavy load; then, with a deep note like the twang of a guitar-string, launches out into the sun-steeped air. But it is all a vain pretence, and well she knows it. Watch her as she flies, and you will see her busy ding-dong pace slacken a dozen yards away. She fetches a turn or two above the leafless apple-branches of the garden, with the rest of the chanting, workless crew. She may presently start off again at a livelier speed than ever, as though vexed at being allured, even for a moment, from the duty that calls her away to the mist-clad hill. But it always ends in the same fashion. A little later she is fluttering down on the threshold of the silent hive, and running busily in, keeping up the transparent fiction, you see, to the last.

An Officious Dame

Many more set themselves to look for sweets where they must know there is little likelihood of finding any. Scarce one goes near the glowing belt of pompons rimming the garden on every side. But here is one bee, an ancient dame, with ragged wings and shiny thorax, poised outside a cranny in the old brick wall, and examining it with serious, shrill inquiry. She is obviously making-believe, to while away the time, that it is a choice blossom full of nectar. She knows it is nothing of the kind; but that will neither check her ardour nor expedite the piece of play-acting. She spins it out to the utmost, and leaves the one dusty crevice at last only to go through the same performance at the next.

I often wonder wherein lies the fascination to a hive-bee of an open window or door. Sitting here ledgering in the little office of the bee-farm—where no honey, nor the smell of honey, is ever allowed to come—sooner or later, in the quiet of the golden morning, the familiar voice peals out. It is startling at first, unless you are well used to it— this sudden high-pitched clamour breaking the silence about you; and the oldest bee-man must lay down pen or rule, and look up from his work to scan the intruder.

She has darted in at the door, and has stopped in mid-air a foot or two within the room. The sound she makes is very different from that of a bee in ordinary flight. You cannot mistake its meaning; it is one long-drawn-out, musical note of exclamation, an intense, reiterated wonder at all about her—the subdued light, the walls covered with book-shelves, the littered table, and the vast wingless, drab-coloured creature sitting in the midst of it all, like a funnel-spider in his snare. Bees entering a room in this way seldom stop more than a second or two, and, more rarely still, alight. As a rule, they are gone the next moment as swiftly as they came, leaving the impression that their quick retreat was due to a sudden accession of fear; just as children, venturing into some dark unwonted place, at first boldly enough, will suddenly turn tail and flee, with terror hard upon their heels.

But what should bring bees into such unlikely situations during these warm bright breaks in the wintry weather, when they seldom or never venture out of the range of hives and fields in the season of plenty? It would be curious to know whether people who have never kept bees, nor handled hives, are habitually pried upon in this way; or whether it is only among bee-men the thing occurs. Naturalists are commonly agreed that bees possess an extraordinary sense of smell; indeed, the fact is patent to all who know anything of hive-life. Now, years of stinging render the bee-master immune to the ordinary results of a prod from a bee's acid-charged stiletto. There is only a sharp prick, a little irritation at the moment, but seldom any after-effects of swelling or inflammation, local or general. But all this injection of formic acid under the skin year after year might very well have a cumulative effect, so that the much-stung bee-man would eventually acquire in his own person the permanent odour of the hive. And this, scented afar off, may well be the attraction that brings these roving scrutineers to places having, in themselves, no sort of interest to the winged hive-people.

The Perils of " Immunity "

The mention of stinging brings back a thought that has often occurred to me. Do lovers of honey ever quite realise the price that must be paid before their favourite sweet is there for them on the break-fast- able, filling the room with the mingled perfume from whole countryside? It is easy to talk of immunit from the effect of bee-stings; but the truth is that this immunity means, for the bee-master, no more than power to go on with his work in spite of the stinging. And this power is not a permanent one. It is brought about by incessant pricks from the living poisoned needle; the ordeal must be continuous, or the immunity will soon pass away. Over-care in handling bees is good only up to a certain point. The bee-man who, by continual practice, has brought this gentlest art to its highest perfection, so that he can do what he likes with his own bees without fear of harm, has, in a sense, created for himself a kind of fools' paradise. All the time his once dear-bought privilege is slowly forsaking him. He is like the Listerist faddist, who so destroys all disease germs in his vicinity that his natural disease-resisting organisation becomes atrophied through want of work. Then, perhaps, his precautions are upheld for a season, whereupon a particularly virulent microbe happens by; and, finding the house empty, swept, and garnished, calls in the seven devils with a will.

Such a contingency is always in wait for the stay-at-home, never-stung bee-master of neighbourly proclivities. Sooner or later he will be called to help some maladroit in bee-craft, whose bees have been thoroughly vitiated by years of "monkeying." And then the rod will come out of pickle to a lively tune. Of course, a little stinging is nothing; but there is no doubt that, with anything over a dozen stings or so at a time, the most hardened and experienced bee-man may easily stand, for a minute or two at least, in danger of losing his life.

So it happened to me once. I had gone to look at a neighbour's stocks. The bees were as quiet as lambs until I came to the seventh hive; and then, with hardly a note of warning, they set upon me like a pack of flying bull-dogs. It is long enough ago now, but I can still give a pretty accurate account of the symptoms of acute formic-acid poisoning. It began with a curious pricking and burning over the entire inner surface of the mouth and throat. This rapidly spread, until my whole body seemed on fire, and the target, as it were, for millions of red-hot darts. Then first my tongue and lips, and every other part of head and neck, in quick succession, began to swell. My eyes felt as though they were being driven out of my head. My breathing machinery seized up, and all but stopped. A giddy congestion of brain followed. Finally, sight and hearing failed, and then almost consciousness.

I can just remember crawling away, and thrusting head and shoulders deep into a thick lilac bush, where the bees ceased to molest me. But it was a good hour or more before I could hold the smoker straight again, and get on with the next stock.

CHAPTER XXVI

THE LONG NIGHT IN THE HIVE

THERE are few things more mystifying to the student of bee-life than the way in which winter is passed in the hive. Probably nineteen out of every twenty people, who take a merely theoretical interest in the subject, entertain no doubt on the matter. Bees hibernate, they will tell you—pass the winter in a state of torpor, just as many other insects, reptiles, and animals have been proved to do. And, though the truth forces itself upon scientific investigators that there is no such thing as hibernation, in the accepted sense of the word, among hive-bees, the perplexing part of the whole question is that, as far as modern observers understand it, the honey-bee ought to hibernate, even if, as a matter of fact, she does not.

For consider what a world of trouble would be saved if, at the coming of winter, the worker-bees merely got together in a compact cluster in their warm nook, with the queen in their midst; and thenceforward slept the long cold months away, until the hot March sun struck into them with the tidings that the willows—first caterers for the year's winged myriads—were in golden flower once more; and there was nothing to do but rouse, and take their fill. It would revolutionise the whole aspect of bee-life, and, to all appearances, vastly for the better. There would be no more need to labour through the summer days, laying up winter stores. Life could become for the honey-bee what it is to most other insects—merry and leisurely. There would be time for dancing in the sunbeams, and long siestas under rose-leaves; and it would be enough if each little worker took home an occasional full honey-sac or two for the babies, instead of wearing out nerve and body in all that desperate toiling to and fro.

Yet, for some inscrutable reason, the honey-bee elects to keep awake—uselessly awake, it seems— throughout the four months or so during which out-door work is impossible; and to this apparently undesirable, unprofitable end, she sacrifices all that makes such a life as hers worth the living from a human point of view.

Restlessness, and the Reason for It

You can, however, seldom look at wild Nature's ways from the human standpoint without danger of postulating too much, or, worse still, leaving some vital, though invisible thing out of the argument. And this latter, on a little farther consideration, proves to be what we are now doing. Prolonged study of hive-life in winter will reveal one hitherto unsuspected fact. At this time, far from settling down into a life of sleepy inactivity, the queen-bee seems to develop a restlessness and impatience not to be observed in her at any other season. It is clear that the workers would lie quiet enough, if they had only themselves to consider. They collect in a dense mass between the central combs of the hive, the outer members of the company just keeping in touch with the nearest honey-cells. These cells are broached by the furthermost bees, and the food is distributed from tongue to tongue. As the nearest store-cells are emptied, the whole concourse moves on, the compacted crowd of bees thus journeying over the comb at a pace which is steady yet inconceivably slow.

But this policy seems in no way to commend itself to the queen. Whenever you look into the hive, even on the coldest winter's day, she is generally alert and stirring, keeping the worker-bees about her in a constant state of wakefulness and care. Though she has long since ceased to lay, she is always prying about the comb, looking apparently for empty cells wherein to lay eggs, after her summer habit. Night or day, she seems always in this unresting state of mind, and the work of getting their queen through the winter season is evidently a continual source of worry to the members of the colony. Altogether, the most logical inference to be drawn from any prolonged and careful investigation of hive-life in winter is that the queen-bee herself is the main obstacle to any system of hibernation being adopted in the hive. This lying-by for the cold weather, however desirable and practicable it may be for the great army of workers, is obviously dead against the natural instincts of the queen. And since, being awake, she must be incessantly watched and fed and cared for, it follows that the whole colony must wake with her, or at least as many as are necessary to keep her nourished and preserved from harm.

The Queen a Slave to Tradition

Those, however, who are familiar with the resourceful nature of the honey-bee might expect her to effect an ingenious compromise in these as in all other circumstances; and the facts seem to point to such a compromise. It is not easy to be sure of anything when watching the winter cluster in a hive, for the bees lie so close that inspection becomes at times almost futile. But one thing at least is certain. The brood-combs between which the cluster forms are not merely covered by bees. Into every cell in the comb some bee has crept, head first, and lies there quite motionless. This attitude is also common at other times of the year, and there is little doubt that the tired worker-bees do rest, and probably sleep, thus, whenever an empty cell is available. But now almost the entire range of brood-cells is filled with resting bees, like sailors asleep in the bunks of a forecastle; and it is not unreasonable to suppose that each unit in the cluster alternately watches with the queen, or takes her " watch below " in the comb-cells.

That there should be in this matter of wintering so sharp a divergence between the instincts of the queen-mother and her children is in no way surprising, when we recollect how entirely they differ on almost all other points. How this fundamental difference has come about in the course of ages of bee-life is too long a story for these pages. It has been fully dealt with in an earlier volume by the same writer—" The Lore of the Honey-Bee "—and to this the reader is referred. But the fact is pretty generally admitted that, while the little worker-bee is a creature specially evolved to suit a unique environment, the mother-bee remains practically identical with the mother-bees of untold ages back. She retains many of the instincts of the race as it existed under tropic conditions, when there was no alternation of hot and cold seasons; and hence her complete inability to understand, and consequent rebellion against the needs of modern times.

The Future Evolution of the Hive

Whether the worker-bees will ever teach her to conform to the changed conditions is an interesting problem. We know how they have " improved " life in the hive—how a matriarchal system of government has been established there, the duty of motherhood relegated to one in the thirty thousand or so, and how the males are suffered to live only so long as their procreative powers are useful to the community. It is little likely that the omnipotent worker-bee will stop here. Failing the eventual production of a queen-bee who can be put to sleep for the winter, they may devise means of getting rid of her in the same way as they disburden themselves of the drones. In some future age the mother-bee may be ruthlessly slaughtered at the end of each season, another queen being raised when breeding-time again comes round. Then, no doubt, honey-bees would hibernate, as do so many other creatures of the wilds; and the necessity for all that frantic labour throughout the summer days be obviated.

This is by no means so fantastic a notion as it appears. Ingenious as is the worker-bee, there is one thing that the mere man-scientist of to-day could teach her. At present, her system of queen-production is to construct a very large cell, four or five times as large as that in which the common worker is raised. Into this cell, at an early stage in its construction, the old queen is induced to deposit an egg; or the workers themselves may furnish it with an egg previously laid elsewhere; or again—as sometimes happens—the large cell may be erected over the site of an ordinary worker-cell already containing a fertile ovum. This egg in no way differs from that producing the common, under-sized, sex-atrophied worker-bee; but by dint of super-feeding on a specially rich diet, and unlimited space wherein to develop, the young grub eventually grows into a queen-bee, with all the queen's extra-ordinary attributes. A queen may be, and often is, raised by the workers from a grub instead of an egg. The grub is enclosed in, or possibly in some cases transferred to, the queen-cell; and, providing it is not more than three days old, this grub will also become a fully developed queen-bee.

Hibernation, and no Honey

But, thus far in the history of bee-life, it has been impossible for a hive to re-queen itself unless a newly-laid egg, or very young larva, has been available for the purpose. Hibernation without a queen is, therefore, in the present stage of honey-bee wisdom, unattainable, because there would be neither egg nor grub to work from in the spring, when another queen-mother was needed, and the stock must inevitably perish. Here, however, the scientific bee-master could give his colonies an invaluable hint, though greatly to his own dis advantage. In the ordinary heat of the brood-chamber an egg takes about three days to hatch, but it has been ascertained that a sudden fall in temperature will often delay this process. The germ of life in all eggs is notoriously hardy; and it is conceivable that by a system of cold storage, as carefully studied and ingeniously regulated as are most other affairs of the hive, the bees might succeed in preserving eggs throughout the winter in a state of suspended, but not irresuscitable life. And if ever the honey-bee, in some future age, discovers this possibility, she will infallibly become a true hibernating insect, and join the ranks of the summer loiterers and merry-makers. But the bee-master will get no more honey.

CHAPTER XXVII

THE AUTOCRAT OF THE BEE-GARDEN

"**B**OOKS," said the Bee-Master of Warrilow, looking round through grey wreaths of tobacco-smoke at his crowded shelves, " books seem to tell ye most things ne'ersome-matter; but when it comes to books on bees—well, 'tis somehow quite another pair o' shoes."

He stopped to listen to the wind, blowing great guns outside in the winter darkness. The little cottage seemed to crouch and shudder beneath the blast, and the rain drove against the lattice-windows with a sobbing, timorous note. The bee-master drew the old oak settle nearer to the fire, and sat for a moment silently watching the comfortable blaze.

" ' True as print,' " he went on, lapsing more and more into the quaint, tangy Sussex dialect, as his theme impressed him; " 'twas an old saying o' my father's; and right enough, maybe, in his time. A' couldn't read, to be sure; so a' might have been ower unsceptical. But books was too expensive in those days to put many lies into."

He took down at random from the case on the

chimney-breast about a dozen modern, paper-covered treatises on bee-keeping, and threw them, rather contemptuously, on the table.

" I'm not saying, mind ye," he hastened to add, " that there's a word against truth in any one of them. They're all true enough, no doubt, for they contradict each other at every turn. 'Tis as if one man said roses was white; and another said, ' No, you're wrong, they're yaller '; and a third said, ' Y'are both wrong, they're red.' And when folks are in dispute in this way, because they agree, and not because they differ, there's little hope of ever pacifying them.

" I heard tell once of a woman bee-keeper years ago, that had a good word about bees. Said she, ' They never do anything invariably '; and she warn't far off the truth. She knew her own sex, did wise Mrs Tupper. Now, the trouble with the book-writers on bees is that they try to make a science of something that can never rightly be a science at all. They try to add two numbers together that they don't know, an' that are allers changing, and are surprised if they don't arrive at an exact total. There's the bees, and there's the weather: together the result will be so many pounds of honey. If the English climate went by the calendar, and the bees worked according to unchangeable rules, you might reckon out your honey-take within a spoonful, and bee-keeping would be little more than sitting in a summer-house and figuring on a slate. But with frosts in June, and August weather in February, and your honey-makers naught but a tribe of whimsy, sex-thwarted wimmin-folk, a nation of everlasting spinsters—how can bee-keeping be anything else than a kind of walking-tower in a furrin land, when every twist an' turn o' the way shows something cur'ous or different? "

He stopped to recharge his pipe from the earthen tobacco-jar, shaped like an old straw bee-hive, which had yielded solace to many a past generation of the Warrilow clan.

" 'Tis just this matter of sex," he continued, " that these book-writing bee-masters seem to leave altogether out of their reckoning. And yet it lies well to the heart of the whole business. In an average prosperous hive there are about thirty thousand of these little stunted, quick-witted worker-bees, not one of which but could have grown into a fully-developed mother-bee, twice the size, and laying her thousands of eggs a day, if only her early bringings-up had been different. But nature has doomed her to be an old maid from her very cradle, although she is born with all the instincts and capabilities for motherhood that you wonder at in a fully grown, prolific queen. And yet the bee-masters expect her to accept her fate without a murmur; to live and work to-day just as she did yesterday and the day before; to tend and feed patiently the young bees that she has been denied all part in producing; to support a lot of lazy drones in luxury and idleness; and generally to act like a reasonable, contented, happy creature all the way through."

He took three or four long, contemplative pulls at his Broseley clay, then came back to his subject and his dialect together.

" 'Tis no wonder," said he, " that the little

worker-bee gets crotchety time an' again. Wimmin-creeturs is all of much the same kidney, whether 'tis bees or humans. Their natur' is not to look ahead, but just to do the next thing. They sees sideways mostly, like a horse with an eye-shade but no blinkers. But now and then they ups and looks straight afore 'em, and then 'tis trouble brewing fer masters o' all kinds, whether in hives or homes o' men. Lot's wife, she were a kind o' bee-woman; and so were Eve. I'd ha' been glad to ha' knowed 'em both, bless 'em! The world 'ud be all the sweeter fer a few more like they. Harm done through being too much of a woman-creetur is never all harm in the long run, depend on't."

With his great sunburnt hand he stirred the flimsy, dog-eared pamphlets about thoughtfully, as a man will stir leaves with a stick.

"Now, 'tis just this way with bees," he went on. "If you study how to keep 'em busy, with plain, right-down necessity hard at their heels, all goes well. The bees have no time for anything but work. As the supers fill with honey you take them off and put empty ones in their place. The queen below fills comb after comb with eggs, and you make the brood-nest larger and larger. There is allers more room everywhere, dropped down from the skies, like; no matter how fast the stock increases, nor how much the bees bring in. Just their plain day's work is enough, and more'n enough, for the best of them. And so the summer heat goes by; the honey harvest is ended; and the bees have had no chance to dwell upon, and grow rebellious over, the wise wrong that nature has done their sex. In bee-life 'tis always evil that's wrought, not by want o' thought, but by too much of it. Bad beemanship is just giving bees time to think."

" Many's the time," continued the bee-master, thrusting the bowl of his empty pipe into the heart of the wood-embers for lustration, and taking a clean one down for immediate use from the rack over his head; " many's the time an' oft it has come ower me that perhaps bees warn't allers as we see them now. Maybe, way back in the times when England was a tropic country, tens of thousands o' years ago, there was no call for them to live packed together in one dark chamber, as they do to-day. If the year was warm all the twelve months through, and flowers allers blooming, there 'ud be no need fer a winter-larder, nor fer any hives at all. Like as not each woman-bee lived by herself then, in some dry nook or other; made her little nest of comb, and brought up her own children, happy and comfortable. Maybe, even—and I can well believe it of her, knowing her natur' as I do—she kept a gurt, buzzing, blusterous drone about the place an' let him eat and drink in idleness while she did all the work, willing enough, for the two. Then, as the world slowly cooled down through the centuries, there came a short time in each year when the flowers ceased to bloom, and the bees found they had to put by a store of honey, to last till the heat and the blossoms showed up again. And there was another thing they must have found out when the cold spell was over the earth. Bees that kept apart by themselves died of cold, but those that huddled together in crowds lived warm enough throughout the winter. The more there were of 'em the warmer they kept, and the less food they needed. And so, as the winters got longer and colder, the bee-colonies increased, until at last, from force of habit, they took to keeping together all the year round. So you see, like as not, 'tis experience as has brought 'em to build their cities of to-day, just as experience, or the One ye never mention, has put the same thing into the hearts o' men."

A sudden flaw of wind struck the little cottage with a sound like thunder, and made the cut-glass lustres on the mantle tinkle and glitter in the yellow candle-glow. The old bee-man stopped, with his pipe half-way to his mouth, nodded gravely towards the window, in a kind of obeisance to the elements, and then resumed his theme.

" But there's a many things about bees," he said, "that no man 'ull come to the rights of, until all airthly things is made clear in the Day o' Days. The great trouble and hindrance to bee-keeping is the swarm, and a good bee-master nowadays tries all he can to circumvent it. But the old habit comes back again and again, and often with stocks of bees that haven't had a fit o' it for years. Now, did ye ever think what swarming must have been in the beginning? "

He suddenly levelled the pipe-stem straight at my head.

" Well, 'tis all speckilation, but here's my idee o' it, for what 'tis worth. Take the wapses: they're thousands of years behind the honey-bee in development, and so they give ye a look, so to speak, into the past. The end of a wapse-colony comes when the females are ready in November; and hundreds of them go off to hide for the winter, each in some hole or crevice, until, in the warm spring days, each comes out to start a new and separate home. Well, perhaps the honey-bees did much the same thing long ago, when they were all mother-bees, in the time when the world was young. And perhaps the swarm-fever in a hive to-day is naught but a kind o' memory of this, still working, though its main use is gone. The books here will tell ye o' many other things brought about by swarming, right an' good enough with the old-fashioned hives. Yet that gainsays nothing. Nature allers works double an' treble handed in all her dealings. Her every stroke tells far and wide, like the thousand ripples you make when you pitch a stone in a pond."

CHAPTER XXVIII

HONEY-CRAFT OLD AND NEW

THERE never comes, in early April, that first bright hot day which means the beginning of outdoor work on the bee-farm, but I fall to thinking of old times with a great longing to have them back again.

Modern beemanship, at least to the wide-awake folk in the craft, brings in gold pieces now where formerly one had much ado to make shillings. But profit cannot always be reckoned in money. The old mysteries and the old delusions were a sort of capital that paid cent per cent if you only humoured them aright. Bee-men, who flourished when there was a young queen upon the throne, wore their ignorance as the parson his silk and lawn. It was something that set them apart and above their neighbours. All that the bees did was put to their credit, just for the trouble of a wise wag of the head and a little timely reticence. The organ-blower worked in full view of the congregation, while the player sat invisibly within, so the blower, after the common trend of earthly affairs, got all the glory for the tune.

196

There are no mysteries now in honey-craft. Science has dragooned the fairies out of sight and hearing as a man treads out sparks in the whin. But, though the mysteries have gone, the old music of the hives is still here as sweet as ever. This morning, when the sun was but an hour over the hilltop, I rose from my bed, and, coming down the creaking stair through the silence and half-darkness, threw the heavy old house-door back. At once the level sunshine and the song of bees and birds came pouring in together. There was the loud humming of bees in the leafing honeysuckle of the porch, and the soft low note of the hives beyond. In its plan to-day Warrilow Bee-farm reveals the whole story of its growth from times long gone to the present. All the hives near the cottage are old-fashioned skeps of straw, covered in with three sticks and a hackle. A little way down the slope the ancient bee-boxes begin, eight-sided Stewartons mostly, with the green veneer of decades upon some of them. Beyond these stand the first rack-frame hives that ever came to Warrilow; and thence, stretching away down the sunny hillside in long trim rows, are the modern frame-bar hives, spick and span in their new Joseph's coats of paint, with the gillyflowers driving golden shafts between them, until they reach the line of sheds—comb and honey-stores, extracting-house, and workshops—marking the distant lane-side.

The Water-carriers

As I stood in the doorway, caught by the mesmeric sheen of the light and the beauty of the morning, the humming of the bees overhead grew louder and louder. There were no flowers as yet to attract them, but in early April the dense canopy of honey-suckle here is always besieged with bees, directly the sun has warmed the clinging dewdrops. These were the water-carriers from the hives. Water at this time is one of the main necessities of bee-life. With it the workers are able to reduce the thick honey and the dry pollen to the right consistency for con-sumption, and can then generate the bee-milk with which the young larvæ are fed. Later on in the day the water-fetchers will crowd in hundreds to the oozy pond-side down in the valley—every bee-garden has its ancestral drinking-place invariably resorted to year after year. But thus early the pond-water is too cold for safe transport by so chilly a mortal as the little worker-bee; so Nature warms a tem-porary supply for her here where the dew trembles like drops of molten rainbow at the tip of each woodbine leaf.

I drank myself a deep draught from the well that goes down a sheer sixty feet into the virgin chalk of the hillside, and fell to loitering through the garden ways. Though it was so early, the little oil-engine down below in the hive-making shed was already coughing shrilly through its vent-pipe, and the saw thrumming. Here and there among the hives my men stooped at their work. The pony was harnessing to the cart, and would soon be plodding the three-mile-long road to the station with the day's deliveries of honey. By all laws of duty I should be down there, taking my row of hives with the rest—master and men side by side like a string of turnip-hoers—busy at the spring examination which, as all bee-men know, is the most important work of the year. But the very thought of opening hives, now in the first warm break of April weather or at any time, filled me with a strange loathing. So it never used to be, never could be, in the old days whose memory always comes flooding back to me at this season with such a clear call and such a hindrance to progress and duty. Then I had as little dreamed of opening a hive as opening a vein. I should have done no more than I was doing now—passing from one old straw skep to another through the sweet vernal sunshine, my boots scattering the dew from the grass as I went, and looking for signs that tell the bee-man nearly all he really needs to know. I shut my ears to the throaty song of the engine. I heard the cart drive away without a thought of scanning its load. I got me down in a little nook of red currant flowers under the wall, where the old straw hives were thickest, and gave myself up to idle dreams, dreams of the bees and bee-men of long ago.

I should be splitting elder, thought I; splitting the long, straight wands to make feeding-troughs. I called to mind doing it, here on this self-same bench near upon fifty years ago, with my father, the woodman, sitting at my elbow learning me. We split the wands clean and true, scooped out the pith from each half, and dammed up its ends with clay. Then, with a handful of these crescent troughs and a can of syrup, we went the round of the garden together looking for stocks that were short of stores. When we found one, we pushed the hollow slip of elder gently into the hive-entrance as far as it would go, and filled it with syrup, filling it again and again throughout the day as the bees within drank it dry.

The Old Style and the New

A queer figure my father cut in his short grey smock and his long lean bent legs encased in leathern gaiters, legs between which, when I was little, and trotting after him, I had always a fine view of the sky. He was never at fault in his estimate of a hive's prosperity. The rich clear song and steady traffic of a well-to-do bee-nation he knew at once from the anxious note and frantic coming and going of a starvation-threatened hive. It was the tune that told him. Nowadays we just rip the coverings from a hive and, lifting the combs out one by one, judge by sheer brute-force of eyesight whether there be need or plenty. " One-thirty-two! "—from my sunny seat under the pink currant blossom I can hear the call of the foreman to the booking 'prentice down in the bee-farm— " One-thirty-two—six frames covered—no moth— medium light—brood over three—mark R.Q." R.Q. means that the stock is to be re-queened at the earliest opportunity. She has been a famous queen in her time—One-thirty-two. This would have been her fourth year, had she kept up her fertility. But " brood over three "—that is to say, only three combs with young bees maturing in them—is not good enough for progressive, up-to-date Warrilow in April, and she must be pinched at last. In the common course, I never let a queen remain at the head of affairs after her second season. Nine out of ten of them break down under the wear and stress of two summers, and fall to useless drone-breeding in the third.

Already the sun has climbed high, and yet I linger, though I know I should be gone an hour ago. The darkness, far away as it seems, will not find all done that should be done on the bee-farm, toil as hard as we may. For these sudden hot days in spring often come singly, and every moment of them is precious. To-morrow the north wind may be keening under an iron-grey sky, and pallid wreaths of snow-flakes weighing down the almond-blossom. So it happened only a year ago, when on the twenty-fifth of April I must clear away the snow from the entrance-boards of the hives. It is, I think, the unending round of business—the itch that is on us now of finding a day's work for every day in the year in modern bee-craft—which has had most to do with the changed times. The old leisure, as well as the old colour and mystery, has gone out of bee-keeping. Between burning-time in August and swarming-time in May there used to be little else for the bee-master to do but smoke his pipe and ruminate and watch the wax flowing into the hives. For we all believed that the little pellets of many-tinted pollen which the bees constantly carry in on their thighs were not food for the grubs in the cells, but wax for the comb-building. I could believe it now, indeed, if I might only sit here long enough; but the busy voices are calling, calling, and I must be gone.

CHAPTER XXIX

THE BEE-MILK MYSTERY

AMONG the innumerable scraps of more or less erroneous information on hive-life, dished up by the popular newspapers in course of the year's round, there is occasionally one which is sure to grip the curious reader's attention. No one expects nowadays to read of the honey-bee without being set agape at the marvellous; but, really, when he is gravely told that the nurse-bees in a hive actually give the breast to their young, suckling them with a secreted liquid which is nothing more or less than milk, the ordinarily faithful newspaper student is entitled to be for once incredulous.

The thing, however, in spite of its grotesque improbability, comes nearer to the plain truth than many another item of bee-life more often encountered and unquestionably accepted. There are veritable nurse-bees in a hive, and these do produce something not unlike milk. In about three days after the egg has been deposited in the comb-cell by the queen, or mother-bee, a tiny white grub emerges. The feeding of this grub is immediately commenced by the bees in charge of the nursery quarters of the hive, and there is administered to it a glistening white substance closely resembling thick cream.

Analysts tell us that this bee-milk, as it is called, is highly nitrogenous in character, and that it has a decidedly acid reaction. It is obviously produced from the mouths of the nurse-bees, and appears to be digested matter thrown up from some part of the bee's internal system, and combined with the secretions from one or more of the four separate sets of glands which open into different parts of the worker-bee's mouth. The power to secrete this bee-milk seems to be normally limited to those workers who are under fourteen or fifteen days old. After that time the bee runs dry, her nursing work is relinquished, and she goes out to forage for nectar and pollen, never, as far as is known, resuming the task of feeding the young grubs. But if the faculty is not exercised, it may be held in abeyance for months together. This takes place at the close of each year, when we know that the last bees born to the hive in autumn are those who supply the milk for the first batches of larvæ raised in the ensuing spring.

It is difficult to keep out the wonder-weaving mood when writing of any phase of hive-life, and especially so when we have this bee-milk under consideration. For all recent studies of the matter tend to prove several facts about it not merely wonderful, but verging on the mysterious.

In the first place, its composition seems to be variable at the will of the bees. The white liquid is supplied to the grubs of worker, queen, and drone, and not only is its nature different with each, but it is even possible that this may be farther modified in the various stages of their development. It is well ascertained that the physical and temperamental differences between queen and worker-bee, widely marked as they appear, are entirely due to treatment and feeding during the larval stage. That the eggs producing the two are identical is proved by the fact that these can be transposed without confounding the original purpose of the hive. The queen-egg placed in the worker-cell develops into a common worker, while the worker-egg, when exalted to a queen's cradle, infallibly produces a fully accoutred queen bee. The experiment can also be made even with the young grubs, provided that these are no more than three days old, and the same result ensues.

A close study of the food administered to bees when in the larval stage of their career is specially interesting, because it gives us the key to many otherwise inexplicable matters connected with hive-life. We do not know, and probably never shall know, how mere variation in diet causes certain organs to appear and certain other bodily parts to absent themselves. If the difference between queen and worker-bee were simply one of development, the worker being only an undersized, semi-atrophied specimen of a queen, there would be little mystery about it. But each has several highly specialised organs, of which the other has no trace, just as each has certain functions reduced to mere rudimentary uselessness, which, in the other, possess enormous development and a corresponding importance.

Clearly the food given in each case has peculiar properties, bringing about certain definite invariable results. We are able, therefore, to say positively that most of the classic marvels of bee-life are built up on this one determined issue, this one logical adjustment of cause and effect. The hive creates thousands of sexless workers and only one fertile mother-bee. It limits the number of its offspring according to the visible food supplies or the needs of the commonwealth. It brings into existence, when necessity calls for them, hundreds of male bees or drones, and when their period of usefulness is over it decrees their extermination. When the queen's fecundity declines, it raises another queen to take her place. It can even, under certain rare conditions of adversity, manufacture what is known as a fertile worker, when some mischance has deprived it of its mother-bee and the materials for providing a legitimate successor to her are not forthcoming. And all these results are primarily brought about by the one means, the one vehicle of mystery—this wonderful bee-milk playing its part at all stages in the honey-bee's life from her cradle to her grave.

For to track down this subtly-compounded elixir through all its various uses one must take a survey of almost the whole round of activities in the hive. The food of the young larvæ, whether of queen or worker, for the first three days after the eggs are hatched, seems to consist entirely of bee-milk. The drone-grub gets an extra day of this richly nitrogenous diet. And for the remaining two days of the grub stage of the bee's life milk is given continuously, but, in the case of the worker and drone, in greatly diminished supply. Its place during these two days is largely taken, it is said, by honey and digested pollen in the worker's instance, and by honey and raw pollen for the males.

The queen-grub alone receives bee-milk, of a specially rich kind and in unlimited quantity, for the whole of her larval life. This "royal jelly," as the old bee-masters termed it, is literally poured into the capacious queen-cell. For the whole five days of her existence as a larva she actually bathes in it up to the eyes. But, as far as is known, she receives no other food during this time. The regular order of her development, and of that of the worker-bee, during the five days of the grub stage has been carefully studied, and it is curious to note that the very time when the queen's special organs of motherhood begin to show themselves coincides exactly with the moment at which the worker-grub's allowance of bee-milk is cut down and other food substituted.

This, no doubt, explains why these organs in the adult worker-bee are so elementary as to be practically non-existent, and accounts for the queen's generous growth in other directions. But it leaves us completely in the dark as to the reason for the worker's subsequent elaboration of such organs as the pollen-carrying device, the so-called wax-pincers, and the wax-secreting glands, of which the queen possesses none. Nor are we able to see how the giving or withholding of the bee-milk should furnish the queen with a long curved sting and the worker with a short straight one; nor how mere manipulation of diet can result in making the two so dissimilar in temperament and mental attributes—the worker laborious, sociable, almost preternaturally alert of mind, and withal essentially a creature of the open air and sunshine; the queen dull of intelligence, possessed of a jealous hatred of her peers, for whom all the light and colour and fragrance of a summer's morning have no allurements, a being whose every instinct keeps her, from year's end to year's end, pent in the crowded tropic gloom of the hive.

But the bee-milk as well as being the main ingredient in the larval food, has other and almost equally important uses. It is supplied by the workers to the adult queen and drones throughout nearly the whole of their lives, and forms an indispensable part of their daily diet. And this gives us a clue in our attempt to understand, not only how the population of the hive is regulated, but why the males are so easily disposed of when the annual drone-massacre sets in. By giving or depriving her of the bee-milk, the workers can either stimulate the queen to an enormous daily output of eggs or reduce her fertility to a bare minimum; and, as for the drones, it is starvation that is the secret of their half-hearted, feeble resistance to fate.

Yet though we may recount these things, and speak of this mysterious essence called bee-milk as really the mainspring of all effort and achievement within the hive, it is doubtful whether we have solved the greatest mystery of all about it. Of what is it composed, and whence is it derived? The generally-accepted explanation of its origin is that it is pollen-chyle regurgitated from the second stomach of the bee, combined with the secretions from certain glands of the mouth in passing. But the most careful dissections have never revealed anything like bee-milk in any part of the bee's internal system. Its pure white, opaque quality has absolutely no counterpart there : nor, indeed—if we are to believe latest investigations—does pollen-chyle exist at all in either the first or second stomach of the bee, whence alone it could be regurgitated. Bee-milk, it would seem, is still a physiological mystery, and so may remain to the end of time.

CHAPTER XXX

THE BEE-BURNERS

COUNTRY wanderings towards the end of summer, even now when the twentieth century is two decades old, still bring to light many ancient and curious things. Within an hour of London, and side by side with the latest agricultural improvements, you can still see corn coming down to the old reaping-hook, still watch the plough-team of bullocks toiling over the hillside, still get that unholy whiff of sulphur in the bee-gardens where the old-fashioned skeppists are "taking up" their bees.

Burning-time came round usually towards the end of August, sooner or later according to the turn of the season. The bee-keeper went the round of his hives, choosing out the heaviest and the lightest stocks. The heaviest hives were taken because they contained most honey; the lightest because, being short of stores, they were unlikely to survive the winter, and had best be put to profit at once for what they were worth. Thus a complete reversal of the doctrine of the survival of the fittest was artificially brought about by the old bee-masters. The most vigorous strains of bees were carefully weeded out year by year, and the perpetuation of the race left to those stocks which had proved themselves malingerers and half-hearts.

There was also another way in which this system worked wholly for the bad. If a hive of bees reached burning-time with a fully charged store-house, it was probably due to the fact that the stock had cast no swarm that year, and had, therefore, preserved its whole force of workers for honey-getting. Under the light of modern knowledge, any stall of bees that showed a lessened tendency towards swarming would be carefully set aside, and used as the mother-hive for future generations; for this habit of swarming, necessary under the old dispensation, is nothing else than a fatal drawback under the new. The scientific bee-master of to-day, with his expanding brood-chambers and his system of supplying his hives artificially with young and prolific queens every third year, has no manner of use for the old swarming-habit. It serves but to break up and hopelessly to weaken his stocks just when he has got them to prime working fettle. Although the honey-bee still clings to this ancient impulse, there is no doubt that selective cultivation will ultimately evolve a race of bees in which the swarming-fever shall have been much abated, if not wholly extinguished; and then the problem of cheap English honey will have been solved. But in ancient times the bee-gardens were replenished only from those hives wherein the swarming-fever was most rampant. The old bee-keepers, in consigning all their heavy stocks to the sulphur-pit, unconsciously did their best to exterminate all non-swarming strains.

The bee-burning took place about sunset, or as soon as the last honey-seekers were home for the night. Small circular pits were dug in some quiet corner hard by. These were about six or eight inches deep, and a handful of old rags that had been dipped in melted brimstone having been put in, the bee-keeper went to fetch the first hive. The whole fell business went through in a strange solemnity and quietude. A knife was gently run round under the edge of the skep, to free it from its stool, and the hive carefully lifted and carried, mouth downwards, towards the sulphur-pit, none of the doomed bees being any the wiser. Then the rag was ignited and the skep lowered over the pit. An angry buzzing broke out as the fumes reached the undermost bees in the cluster, but this quickly died down into silence. In a minute or two every bee had perished, and the pit was ready for the next hive.

That this senseless and wickedly wasteful custom should have been almost universal among bee-men up to comparatively recent times is sufficiently a matter for wonder; but that the practice should still survive in certain country districts to-day wellnigh passes belief. If the art of bee-driving—a simple and easy method by which all the bees in a full hive may be transferred unhurt to an empty one, and that within a few minutes—were a new discovery, the thing might be condoned as all of a piece with the general benightedness of mediæval folk. But bee-driving was known, and openly advocated, by several writers on apiculture at least a hundred years ago. By this method, just as easy as the old and cruel one, not only do the entire stores of each hive fall into the undisputed possession of the bee-master, but he retains the colony of bees complete and unharmed for future service. He has secured all the golden eggs, and the goose is still alive.

Those who desire to make a start in beemanship inexpensively might do worse than adopt a practice which the writer has followed for many years past. As soon as the time for the bee-burners' work arrives, a bicycle is rigged up with a bamboo elongation fore and aft. From this depend a number of straw skeps tied over with cheese-cloth. A bee-smoker and a set of driving-irons complete the equipment, and there is no more to do than sally forth into the country in search of condemned bees.

It is usually not difficult to persuade the cottage apiarist to let you operate on his hives. As soon as he learns that all you ask for your trouble is the bees, while you undertake to leave him the entire honey-crop and a *pour-boire* into the bargain, he readily gives you access to his stalls. The work before you is now surprisingly simple. A few strong puffs of smoke into the entrance of the hive under manipulation will effectually subdue the bees. Then the hive is lifted, turned over, and placed mouth upwards in any convenient receptacle—a pail or bucket will do, and will hold it as firmly as need be. Your own travelling-gear now comes into use. One of the empty skeps is fitted over the inverted hive. The two are pinned together with an ordinary meat-skewer at one point, and then the skep is prised up and fixed on each side with the driving-irons, so that the whole looks like a box with the lid half-raised. Now you have merely to take up a position in front of the two hives, and begin a steady gentle thumping on the lower one with the palms of the hands.

At first, as the combs begin to vibrate, nothing but chaos and bewilderment are observable among the bees. For a moment or two they run hither and thither in obvious confusion. But presently they seem to get an inkling of what is required of them, and then follows one of the most interesting, not to say fascinating, sights in the whole domain of bee-craft. Evidently the bees arrive at a common agreement that the foundations of their old home have become, from some mysterious cause or other, undermined and perilous; and the word goes forth that the stronghold must be abandoned without more ado. On what initiation the manœuvre is started has never been properly ascertained; but in a little while an ordered discipline seems to spread throughout the erstwhile distracted multitude. In one solid hurrying phalanx the bees begin to sweep up into the empty skep. Once fairly on the march, the process is soon completed. In eight or ten minutes at most, the entire colony hangs in a dense compact cluster from the roof of your hive. Below, brood-combs and honey-combs are alike entirely deserted. There is nothing left for you to do now but carefully to detach the upper-most skep: replace the cheese-cloth, thus securing your prisoners for their journey to their new home; and to set about driving the next stock.

CHAPTER XXXI

EVOLUTION OF THE MODERN HIVE

THE bee-master, explaining to an interested novice the wonders of the modern bar-frame hive, often finds himself confronted by a very awkward question. He is at no loss for words, so long as he confines himself to an enumeration of the hive's many advantages over the ancient straw skep—its elastic brood and honey chambers, its movable combs interchangeable with all other hives in the garden, its power of doubling and trebling both the number of worker-bees in a colony and the amount of harvested honey; above all, its control over sanitation and the breeding of unnecessary drones. But when he is asked the question: Who invented this hive which has brought about such a revolution in bee-craft? his eloquence generally comes to a dead stop. Perhaps one in a hundred of skilled modern bee-keepers is able to answer the query. But the ninety-nine will tell you the bar-frame hive had no single inventor; it came to its latter-day perfection by little and little—the conglomerate result of years of experience and the working of many minds.

214

This is, of course, as true of the modern bee-hive as it is of all other appliances of world-wide utility. But it is equally true that everything must have had a prime inception at some time, and through some special human agency or other; and, in the case of the bar-frame hive, the honours appear to be pretty equally divided between two personages widely separated in the world's history—Samson and Sir Christopher Wren.

Perhaps these two names have never before been bracketed together either in or out of print; yet that the association is not a fanciful, but in all respects a natural and necessary one will not be difficult to prove.

The story of how Samson, albeit unconsciously, first gave the idea of the movable comb-frame to an English bee-master is probably new to most apiarians. As to whether the cloud of insects which Samson saw about the carcase of the dead lion were honey-bees or merely drone-flies, we need not here pause to determine. We are concerned for the moment only with one modern explanation of the incident. This is that, although honey-bees abominate carrion in general, in this particular case the carcase had been so dried and emptied and purified by the sun and usual scavenging agencies of the desert as to leave nothing but a shell—a very serviceable makeshift for a bee-hive, in fact—consisting of the tanned skin stretched over the ribs of the lion.

In the summer of 1834 a certain Major Munn was walking among his hives, pondering the ancient Bible narrative, when a sudden brilliant idea occurred to him. Like most advanced bee-keepers of his day, he had long grown dissatisfied with the straw hive, and his bees were housed in square wooden boxes. But these, although more lasting, were nearly as unmanageable as the skeps. The bees built their combs within them on just the same haphazard plan; and, once built, the combs were fixed permanently to the tops of the boxes. Now, the idea which had occurred to Major Munn was simply this: He reflected that the combs built by the bees in the dry shell of the lion-skin were probably attached each to one of the encircling ribs; so that, when Samson took the honey-comb, all he need have done was to remove a rib, bringing the attached comb away with it. Thereupon Major Munn set to work to make a hive on the rib-plan, which was composed of a number of wooden frames standing side by side, each to contain a comb and each removable at will. Since that time numberless small and great improvements have been devised; but, in its essence, the modern hive is no more than the dried lion-skin distended by the ribs, as Samson found it on that day when he went on his fateful mission of wooing.

The part played by Sir Christopher Wren in the evolution of the bar-frame hive, though not so romantic, was fraught with almost equal significance to modern bee-craft. Movable comb-frames were as yet undreamed of in Wren's time, nearly two hundred years before Major Munn invented them. But Wren seems to have been the discoverer of a principle just as important. This was what latter-day bee-keepers call "storification." Wren's hive consisted of a series of wooden boxes, octagonal in shape, placed one below the other, with intercommunicating doors, and glass windows in the sides of each section. Up to that date bee-hives had been merely single receptacles made of straw, plastered wattles, or wood. When the stock had outgrown its dwelling there was nothing for it but to swarm. But by the device of adding another story below the first one, when this was crowded with bees, and a third or even a fourth if necessary, Wren was able to make his hive grow with the growth of his bee-colony or contract with its post-seasonal decline. He had, in fact, invented the elastic brood-chamber, which alone enables the bee-master to put in practice the one cardinal maxim of successful bee-keeping—the production of strong stocks.

Wren's octagon storifying hive seems to have been plagiarised by most eminent bee-masters of his day and after with the naive dishonesty so character-istic among bee-men of the time. Thorley's hive is obviously taken from, indeed, is probably identical with, that of Wren. The hive made and sold by Moses Rusden, King Charles II.'s bee-master, is of almost exactly the same pattern, but it is described as manufactured under the patent of one John Geddie. This patent was taken out by Geddie in 1675, and Geddie would appear to be the arch-purloiner of the whole crew. For it is quite certain that, having had one of Wren's hives shown to him, he was not content with merely copying it, but actually went and patented the principle as his own idea.

But Wren's hive, good as it was in comparison with the single-chambered straw skep or wooden box, still lacked one vital element. Although he and his imitators had realised the advantage of an expanding bee-hive, this was secured only by the process of "nadiring," or adding room below. Thus the upper part of Wren's hive always contained the oldest and dirtiest combs, and as bees almost invariably carry their stores upwards, the production of clear, uncontaminated honey under this system was impossible. It remained for a Scotsman, Robert Kerr, of Stewarton, in Ayrshire, to perfect, some hundred and fifty years later, what Wren had so ingeniously begun.

Whether Kerr—or " Bee Robin," as he was called by his neighbours—ever saw or heard of hives on Sir Christopher Wren's plan has never been ascertained. But plagiarism was in the air throughout those far-off times, and there is no reason to think Kerr better than his fellows. In any case, the " Stewarton " hive, like Wren's, was octagon in shape, and had several stories; but these stories were added above as well as below. By placing his empty boxes first underneath the original brood-chamber, to stimulate increase of population, and then, when the honey-flow began, placing more boxes above to receive the surplus honey, " Bee Robin " succeeded in getting some wonderful harvests. His big supers, full of snow-white virgin honey-comb, were soon the talk of Glasgow, where he readily sold them. Imitators sprang up far and near, and it is only within the last twenty-five or thirty years that his hives can be said to have fallen into desuetude.

But probably his success was due not more to his invention of the expanding honey-chamber than to two other important innovations which he effected in bee-craft. The octagonal boxes of Wren had fixed tops with a central hole, much like the straw hive still used by the old-fashioned bee-keepers to this day. "Bee Robin" did away with these fixed tops, and substituted a number of parallel wooden bars from which the combs were suspended, the spaces between the bars being filled by slides withdrawable at will. He could thus, after having added a story to his honey-chamber, allow the bees access to it by withdrawing his slides from the outside: and when the super was filled with honeycomb, the slides were again employed in shutting off communication, whereupon the super could be easily removed.

This, however, though it greatly facilitated the work of the bee-master, did not account for the large yields of surplus honey, which the "Stewarton" hive first made possible. In the light of modern bee-knowledge, it is plain that a big honey-harvest can only be secured by a corresponding large stock of bees, and Robert Kerr seems to have been the originator of what was nothing less than a revolution in the craft. Hitherto the bee-keeper had estimated his wealth according to the number of his hives, and the more these subdivided by swarming, the more prosperous their owner accounted himself. But "Bee Robin" reversed all this. He housed his swarms not singly, but always two at a time; and he made large stocks out of small ones by the simple expedient of piling the brood-boxes of several colonies together. In a word, it was the "Dreadnought" principle applied to the peaceful traffic of the hives.